Framework
SCIENCE
Foundations

Sarah Jagger

OXFORD
UNIVERSITY PRESS

OXFORD
UNIVERSITY PRESS

Great Clarendon Street, Oxford OX2 6DP

Oxford University Press is a department of the University of Oxford.
It furthers the University's objective of excellence in research, scholarship,
and education by publishing worldwide in

Oxford New York

Auckland Cape Town Dar es Salaam Hong Kong Karachi
Kuala Lumpur Madrid Melbourne Mexico City Nairobi
New Delhi Shanghai Taipei Toronto

With offices in

Argentina Austria Brazil Chile Czech Republic France Greece
Guatemala Hungary Italy Japan South Korea Poland Portugal
Singapore Switzerland Thailand Turkey Ukraine Vietnam

Oxford is a registered trade mark of Oxford University Press
in the UK and in certain other countries

British Library Cataloguing in Publication Data

Data available

ISBN 10: 0 19 915003 6

ISBN 13: 978 019 915003 8

1 3 5 7 9 10 8 6 4 2

Printed in Italy by Rotolito Lombarda

Acknowledgements

The author and publisher would like to thank the following for permission
to reproduce photographs:

p29 Photodisc/OUP; p36 OUP; p37 Paddy Gannon; p38 Andrew Lambert
Photography/Science Photo Library; p46 Photodisc/OUP; p66 Paddy Gannon;
p67 Alex Bartel/Science Photo Library; p68 Zooid Pictures; p78 Photodisc/
OUP; p81 Photodisc/OUP

Technical illustrations by Oxford Designers & Illustrators

Cartoons by John Hallet

Front cover photo: Photodisc and Digital Vision

Contents

7E Acids and alkalis

7F Simple chemical reactions

7G Particle model of solids, liquids and gases

7H Solutions

Introduction

Everyone in school does science because it helps us make decisions about our everyday lives. For example: Which foods should we eat if we want to be healthy? Is it safe to use a mobile phone? Science is also fun, and by studying science we get to find out how things work. This book is designed to help you learn about the key ideas in science that are taught in Year 7. We hope that you enjoy it.

How to use this book

This book is divided into 12 topics:

O At the start of each topic you will find an **opener page**. This page will remind you of what you already know about a topic, and will introduce the key ideas that you are about to meet.

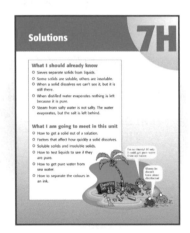

O At the bottom of each main page there are some **questions** for you to test your understanding. Most of them can be answered using the information on the page, but some will require you to use your thinking skills and apply what you have just learnt. These questions are indicated by a thought bubble like this one:

O At the end of each topic there is a '**What have I learnt?**' page, with questions for you to test yourself.

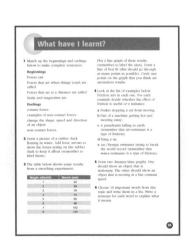

O If you want to find out about something in particular use the **Contents** or the **Index**.

O The **Glossary** explains what certain words mean.

Cells

> **What I should already know**
>
> O Plant and animals are living things.
>
> O Living things grow, reproduce, feed and move.
>
> O Humans have a heart that pumps blood.
>
> O Micro-organisms are very small.
>
> **What I am going to meet in this unit**
>
> O Plants and animals are made up of cells.
>
> O What the parts of a cell do.
>
> O How plants and animals are similar.
>
> O What makes plants different to animals.
>
> O How we grow.
>
> O Why some cells are specialised.
>
> O Tissues and organs.

All living things are made up of **cells**. Cells are sometimes called the building blocks of life. They build living things, like bricks are used to build houses. Cells are so small that they can only be seen through a microscope.

The house in the picture is built from many bricks. We are built from many cells.

Animal cells

An animal cell has three parts:

o The **cell membrane** is a thin skin that controls what goes into and out of the cell.

o The **cytoplasm** is like jelly. It is where important chemical reactions happen.

o The **nucleus** controls what the cell does.

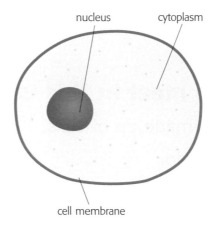

nucleus   cytoplasm

cell membrane

1 Unscramble these words:

lecl barnmeem   **taplomsyc**   **luncsue**

2 Use the words from question **1** to complete this table:

Cell part	What it does
	a thin skin that controls what goes into and out of a cell
	where chemical reactions happen
	controls what the cell does

3 Cells in living organisms are like the bricks in a house. Think of another way to describe what a cell is like. Explain your idea.

Copy and complete using these words:

cell membrane

nucleus   **cells**

cytoplasm

All living things are made up of _____.

Cells contain a nucleus, _____ and _____ _____. The _____ controls what the cell does.

Plant cells

Like animals, plants are also made up of cells. Plant cells have all the cell parts that are found in animal cells. They have a **cell membrane**, **cytoplasm** and a **nucleus**.

Plant cells also have three extra cell parts that animal cells don't have:

o The **cell wall** supports the cell and gives it its shape.

o **Chloroplasts** trap light energy so that the plant can make its own food. They contain chlorophyll (the chemical that makes plants green).

o The **vacuole** contains cell sap and stores important chemicals.

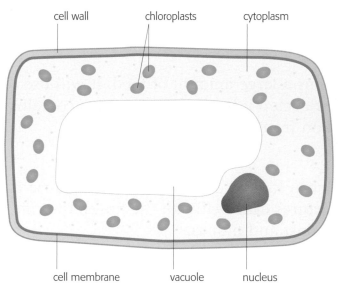

cell wall   chloroplasts   cytoplasm

cell membrane   vacuole   nucleus

1 Which part of a plant cell makes the plant's food?

2 Which of these cell parts are only found in plant cells? Which are found in both plant and animal cells?

**vacuole   cell membrane
nucleus   cell wall**

3 Why do you think animal cells don't have cell walls? Explain your answer.

Copy and complete using these words:

**cell wall   animal
chloroplast
vacuole**

Plant cells have all the cell parts found in _____ cells. They also contain three extra parts. These are the _____, cell membrane and _____. The _____ _____ gives the plant cell its shape.

Are all cells the same?

Some cells look and behave differently to other cells. This helps them to do different jobs. They are called **specialised** cells.

Sperm cell

A sperm cell looks like a tadpole. It has a tail to swim with and its head has a coating to help it get into an egg cell. It also contains genetic information from the father.

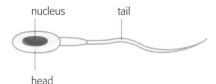

Egg cell

An egg cell contains a large amount of food. It will use this to grow and develop if it is fertilised. It also contains genetic information from the mother.

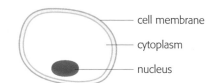

Epithelial cell

An epithelial cell can look like a comb. This is because it has hair-like structures called cilia. In the lungs, the cilia help to get rid of dust and mucus.

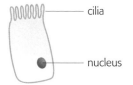

Root hair cell

Root hair cells are found in the roots of plants. They are long and thin, and have a large surface area. They absorb water and minerals from the soil.

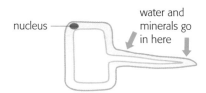

1 What do we call a cell that looks and behaves differently to other cells?

2 These words have had their vowels removed. What should they say?

   rt hr cll   **sprm cll**

3 Choose one of the cells above. Draw a picture of the cell and describe how it is specialised.

Copy and complete using these words:

epithelial   **egg**   **specialised**
root hair   **sperm**

Cells can be _____ to do different jobs. _____ and _____ cells carry genetic information. An _____ cell may have structures called cilia. A _____ _____ cell absorbs water and minerals.

Cell		Living things are made up of cells, like houses are made up of bricks.	
Tissue	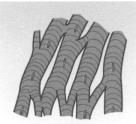	A group of cells of the same type working together makes a tissue, like a group of bricks makes a wall.	
Organ	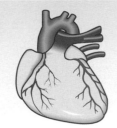	In our bodies different tissues work together to make an organ. In a house different materials are used to make a room.	
Organ system	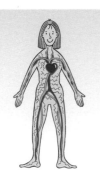	In the body organs that work together make an organ system, like several rooms make a house.	

1 Unscramble these words.

garno   **suseti**

2 Name a type of tissue that is found in the heart.

3 Which of these organs are found in plants and which in animals?

leaf   heart   lung   root

Copy and complete using these words:

organs   system   tissues

_____ are formed from groups of cells of the same type. _____ are made when different tissues work together. The circulatory _____ consists of blood vessels and the heart.

Cell division

Our bodies grow and replace damaged cells by **cell division**. Cells are replaced all the time.

Cell division happens in four stages:

1 The cell **grows** and gets **bigger**.

2 The information in the nucleus is **copied** (**duplicated**).

3 The nucleus in the cell splits into two. Each new nucleus has the same information inside it.

4 When the cell is ready, it splits into two cells. This is called **dividing**.

These two new cells then each divide into two more new cells, which can divide again. This goes on and on to make lots of new cells.

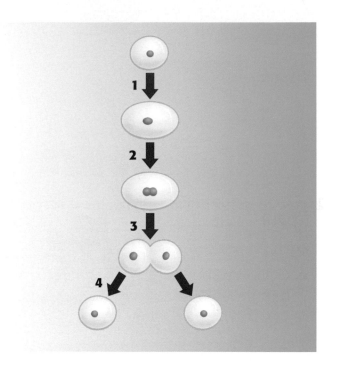

1 What word means the same as 'copied' when we describe what happens to the information in the nucleus during cell division?

2 These words have had their vowels removed. What should they say?

   grw   splt   dvdng

3 Draw a diagram to show how cell division can make eight new cells from just one old cell.

Copy and complete using these words:

copied   cell division   divides information   splits   grows

Our bodies grow and replace damaged cells by _____ _____.

First, the cell _____ and gets bigger before the _____ in the nucleus is _____. The nucleus then _____ into two. After this, the cell _____ into two new cells.

What have I learnt?

1 Copy the diagram below and add the missing labels. Explain what each part of the cell does.

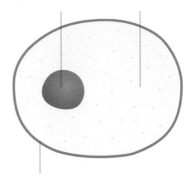

2 Decide whether each statement is true or false. Write them in your book, correcting the ones that are false.

a Cells are sometimes called the building blocks of life.

b Animal cells have both a cytoplasm and a cell wall.

c Animal cells contain all the same parts as plant cells.

3 Match these pictures of specialised cells to their jobs.

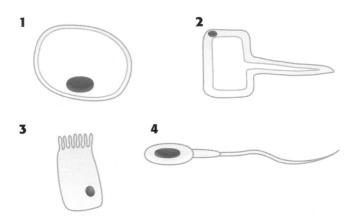

a Contains genetic information and is able to swim.

b Wafts dust and mucus out of airways.

c Takes up water and minerals.

d Contains genetic information and a food store.

4 a Copy and complete the diagram below.

cells ⟶ _____ ⟶ _____ ⟶ organ systems

b Name an organ found in a plant.

c Name an organ that contains muscle cells.

5 Draw and label a diagram to show how new cells are made by cell division.

6 Write a short story to describe a journey through a plant cell. Make it clear what kind of plant cell it is. Describe the different parts of the cell that you see along the way, and what they do.

When you have finished, re-write the story to describe a journey through an animal cell. Which parts of the animal cell did you also see in the plant cell?

Reproduction

What I should already know

○ Humans go through different stages of growth and development.

○ How to describe the differences between new born humans and other animals.

○ Humans and other animals need their parents for different lengths of time.

○ Sperm and egg cells are specialised cells.

○ Reproduction is a life process.

What I am going to meet in this unit

○ Why living organisms reproduce.

○ The different stages in the human lifecycle.

○ The human reproductive organs.

○ Why we aren't identical to our parents.

○ Fertilisation.

○ How a fetus gets its food.

The male reproductive system

New living organisms are made by reproduction. If living organisms did not reproduce they would die out.

Sex cells are made by the **reproductive system**. Men make sex cells called **sperm**. Women make sex cells called **eggs**. To make a new baby, a sperm and an egg must join together.

The male reproductive system

Men and women have different reproductive systems so that they can make different sex cells.

nucleus – containing genetic information

tail – to help it swim

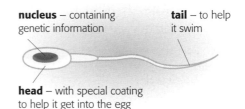

head – with special coating to help it get into the egg

The sperm cell is **adapted** (has special features) to do its job.

sperm tube – the tube the sperm travel along to get from the testes to the penis

glands – where a liquid is added to give the sperm cells energy

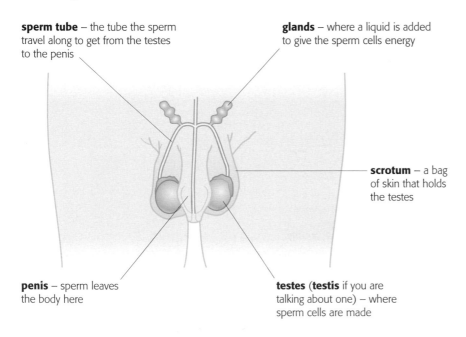

scrotum – a bag of skin that holds the testes

penis – sperm leaves the body here

testes (**testis** if you are talking about one) – where sperm cells are made

1 Where are sperm cells made?

2 These words have had their vowels removed. What should they say?

   scrtm   **sprm tb**   **pns**

3 Describe what these parts of the male reproductive system do:

   scrotum   **glands**   **penis**

Copy and complete using these words:

testes   sperm   heads   tail

Male sex cells are called _____. They have a _____ to help them swim and a special coating on their _____.

Sperm are made in the _____ and travel along the _____ tube to the penis.

The female reproductive system

The female reproductive system

The **female reproductive system** produces sex cells called **eggs**. It also provides a safe environment for a new baby to develop and grow.

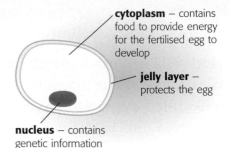

cytoplasm – contains food to provide energy for the fertilised egg to develop

jelly layer – protects the egg

nucleus – contains genetic information

Egg cells are adapted to do their job. They are only about 0.1mm in diameter. That's 0.0001cm!

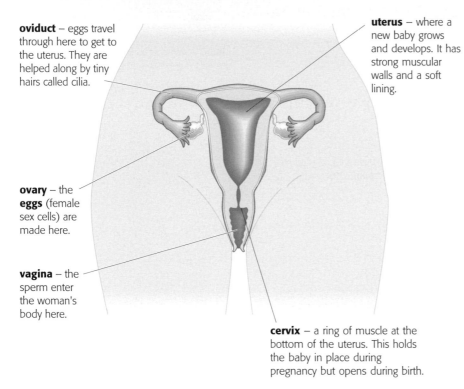

oviduct – eggs travel through here to get to the uterus. They are helped along by tiny hairs called cilia.

uterus – where a new baby grows and develops. It has strong muscular walls and a soft lining.

ovary – the **eggs** (female sex cells) are made here.

vagina – the sperm enter the woman's body here.

cervix – a ring of muscle at the bottom of the uterus. This holds the baby in place during pregnancy but opens during birth.

1 What is the name of the female sex cell?

2 Unscramble these words:

   nagvia   **vecxir**   **seturu**   **rayvo**

3 Which of these parts are found in the female reproductive system?

   ovary   **cervix**   **glands**
   uterus   **testes**   **vagina**

Copy and complete using these words:

oviduct   **ovaries**   **uterus**
vagina   **cervix**

Eggs are made in the _____ and travel down the _____ to the _____, where a new baby can develop. The _____ is a ring of muscle at the bottom of the uterus. The sperm enter the body through the _____.

How the egg and sperm meet

When a man and woman make love, sperm are released from the man's penis into the woman's vagina. The sperm then swim up the oviduct to meet the egg.

If there is no egg in the oviduct the sperm die and no baby is produced. If there is an egg in the oviduct the sperm try to burrow into it.

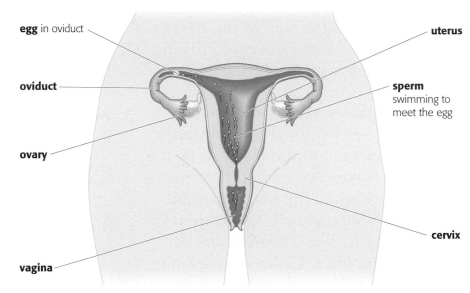

Fertilisation

Once a sperm enters the egg its nucleus fuses with the nucleus of the egg. This is fertilisation. A new baby is made using the genetic information from both the mother and the father. It will have characteristics from both parents.

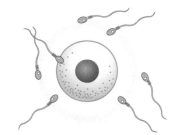

Usually only one sperm fertilises the egg.

1 What happens at fertilisation?

2 Write these sentences in the correct order:
  o The fertilised egg passes into the uterus.
  o An egg is released from the ovary.
  o The egg is fertilised in the oviduct.

3 Why does the egg have to be fertilised inside the mother's body?

Copy and complete using these words:
fuse   fertilisation   oviduct nucleus
To make a baby, an egg cell and a sperm cell must _____ together. The egg and sperm must meet in the _____. The _____ from the sperm joins with the nucleus of the egg. This is called _____.

The developing baby

Once an egg is fertilised, it starts to divide and passes along the **oviduct**. After about one week, it **implants** (sticks) itself into the wall of the **uterus**. It is now called an **embryo**. The embryo keeps growing. After about nine weeks it becomes a **fetus**.

The fetus gets everything it needs from its mother's body through the **placenta**. The fetus is joined to the placenta by the **umbilical cord**. Oxygen and food pass from the mother to the fetus. Harmful substances like nicotine from cigarettes, and alcohol can also pass from the mother to the fetus. Carbon dioxide and other types of waste travel back along the cord to the mother.

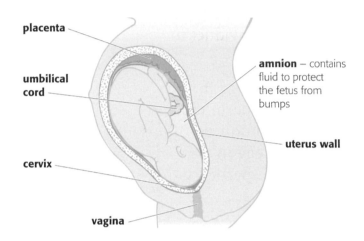

placenta

umbilical cord

cervix

vagina

amnion – contains fluid to protect the fetus from bumps

uterus wall

Birth

After about nine months the baby is ready to be born. It is pushed out of the uterus by muscle **contractions**. The walls of the uterus squeeze the baby out of the mother's body through the vagina.

1 Unscramble these words:

bymore   stufe   altanpce

2 Write a list of things that pass from the mother to the fetus along the umbilical cord. What passes back to the mother from the fetus?

3 How could a developing baby be harmed if its father smokes? Explain your answer.

Copy and complete using these words:

placenta   contract   nine
fetus   uterus

The developing _____ is protected and cared for in the _____, and gets its food and oxygen through the _____. After _____ months the baby is ready to be born. This happens when the muscles of the uterus begin to _____.

Between the ages of 10 and 18 we go through a period of change called **adolescence**. An important part of adolescence is **puberty**. During puberty our bodies go through changes that are controlled by **sex hormones**. We develop into adults that can produce children (we become sexually mature). Girls usually start puberty earlier than boys.

What happens in girls?

o skin becomes oily

o breasts grow

o hair grows under the arms

o ovaries start to release eggs (**menstrual cycle** starts)

o hips get wider

o pubic hair grows

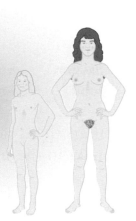

What happens in boys?

o voice deepens

o shoulders widen

o hair grows under the arms, on the face and chest

o muscles grow

o skin becomes oily

o sperm are produced

o pubic hair grows

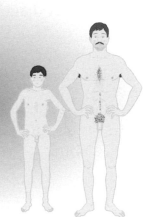

The menstrual cycle

The **menstrual cycle** lasts about 28 days. It starts when a ripe egg is released from one of the ovaries. If the egg is not fertilised it leaves the body with the womb lining. This is called a **period**. The cycle can then start again.

1 These words have had their vowels removed. What should they say?

pbrty   mnstrl cycl   prd

2 What happens during puberty?

3 How do you think your body will change during puberty? Explain why this will happen.

Copy and complete using these words:

sperm   eggs   puberty   menstrual

During adolescence our bodies go through lots of changes. This is called _____. Girls begin to release _____ from their ovaries and the _____ cycle starts. Boys begin to produce _____.

What have I learnt?

1 Copy and complete the sentence below by choosing the correct ending:

Sperm cells are made in the . . .

glands, where a liquid is added.

penis, where they leave the body.

testes, which are held by the scrotum.

2 Copy this diagram into your book. Write each of the sentences below it next to the correct label, to explain what the parts of an egg cell do:

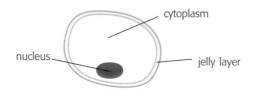

- o contains food to provide energy for the fertilised egg to develop
- o protects the egg
- o contains genetic information

3 Copy these sentences into your book. You will need to unscramble the words in bold so that they are spelt correctly.

a An egg is released from the **royav**.

b The egg passes along the **idtouvc**.

c A sperm meets the egg in the oviduct and **strifilees** it.

d The fertilised egg passes into the **seturu**.

4 Copy this table into your book. Put the bold words below it into the correct columns to show whether they pass from the mother to the fetus through the placenta, or from the fetus to the mother.

Mother → Fetus	Fetus → Mother

oxygen **waste products** **alcohol**
food **carbon dioxide** **nicotine**

5 If an egg is not fertilised, what happens next in the menstrual cycle?

6 Imagine that you are a sperm cell. Draw a comic strip to tell your life story. Your life story should include the answers to these questions:

- o Where were you made?
- o Did you manage to fertilise an egg?
- o If so, how did you get to the egg?

Environment and feeding relationships

What I should already know

- The place where an organism lives is called its habitat.

- How to describe a habitat in terms of the conditions there.

- Different organisms live in different habitats, and eat different types of food.

- Predators hunt and eat prey.

- How to draw a food chain.

What I am going to meet in this unit

- The features of different habitats.

- Organisms are adapted to live in their habitats.

- The environment changes from day to night.

- The environment changes at different times of the year.

- Characteristics of predators and prey.

- How to combine food chains to produce a food web.

Something isn't right!

A **habitat** is a place where plants and animals live. Different habitats have different **features**.

Environmental factors

The **environmental factors** in a habitat determine which organisms will live there. Different environmental factors suit some organisms better than others.

Environmental factors vary from place to place. They include things like:

o how much **light** there is

o how much **oxygen** there is

o the highest and lowest **temperatures**

o how much **water** is available

o the **nutrients** that are available.

Any environment should at least provide animals with food and water, whilst plants need water and light.

Three different habitats.

1 What do we call a place where plants and animals live?

2 These words have had their vowels removed. What should they say?

tmprtr   **wtr**   **lght**   **xygn**

3 Match the three animals below to their habitats.

camel	local pond
polar bear	desert
frog	north pole

Copy and complete using these words:

light   **temperatures**   **habitat**
environmental factors   **organisms**

A place where plants and animals live is called a _____. Different habitats have different _____ _____ that determine which _____ will live there. They include levels of _____ and the highest and lowest _____.

Adapted to survive

Living things have special features that help them **survive** (live) in their habitats. These are called **adaptations**.

Penguin
Penguins have **streamlined** bodies (they can move through water quickly) and are **insulated** (kept warm) by thick layers of fat.

Mole
Moles spend most of their time underground in the dark. They have poor eyesight but are good at **sensing** smells and vibrations. They have short front limbs for digging.

Stick insects
Stick insects live on plants and are usually green or brown. Their thin bodies and legs make them look like twigs. They are **camouflaged** (hard to see).

Tillandsia plants
Tillandsia plants are adapted to grow on the highest branches of trees in the rainforest. They are far from the ground so take up water through their leaves.

1 What do we call the special features that help living things survive in their habitats?

2 How are these organisms adapted to their habitats?

   stick insect   penguin

3 What adaptations might you need if you lived in the sea?

Copy and complete using these words:

adapted   camouflage   fat   streamlining

Organisms are _____ to survive in their habitats. Types of adaptation include _____ so that the organism can't be seen, _____ to move quickly and _____ to insulate the body.

It's all change!

Daily changes

Over 24 hours the conditions in a habitat change. Animals and plants are adapted to cope with these **daily changes**. Examples of daily changes are:

o brightness of the light

o temperature

o humidity (how damp the air is).

Sunflowers follow the Sun as it moves across the sky. This helps them to receive as much **light** as possible.

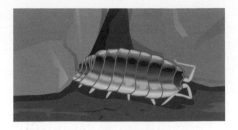

Woodlice are **nocturnal** (active at night) because they prefer cool, damp conditions.

Seasonal changes

Other changes occur throughout the year. In the winter, it is cooler and gets dark much earlier. Plants and animals also have to cope with these **seasonal changes**.

Some birds **migrate** (fly away) in the winter to warmer places where there is more food.

Some animals eat extra food in the autumn so that they can **hibernate** (sleep) through the winter.

Deciduous trees shed their leaves in the winter and live off the food that they made and stored during the summer.

1 Name two daily changes that might affect a habitat.

2 How does a sunflower cope with daily changes in light levels?

3 The fur of an arctic fox turns white in the winter. Why do you think this happens?

Copy and complete using these words:

nocturnal seasonal hibernate daily

_____ changes occur over 24 hours. _____ changes occur over a year. _____ animals come out at night. Animals that _____ sleep through the winter.

Predators and prey

Adapted to eat

Predators are animals that hunt other animals. They are **adapted** to find, catch and kill their **prey**. Some of these adaptations include:

o sharp claws, teeth and talons

o excellent eyesight

o speed (to help them catch their prey).

Lions have sharp claws and teeth to grab and kill their prey.

Owls have large eyes and thick feathers so that they are almost silent when they fly.

Avoiding being eaten

Prey are animals that are hunted and eaten by predators. They also have adaptations, but these are to help them survive and stop them from being eaten!

This moth is **camouflaged**. It is very hard for predators to see it when it is resting on a tree trunk.

A porcupine has quills (sharp spikes). This makes it difficult (and painful) for a predator to eat it.

1 What do we call an animal that is hunted and eaten by other animals?

2 Name three adaptations that predators can have that help them hunt and kill their prey.

3 How is a shark adapted to be a good predator?

Copy and complete using these words:

predators   adapted   prey   survive

Animals are _____ to help them to _____. _____ are animals that hunt and eat other animals. _____ are animals that are eaten by other animals.

Food chains

Food chains show who eats who. They always start with a **producer** (a plant) and finish with a top **consumer**. The arrows show how **energy** is transferred (passed) from one organism to another.

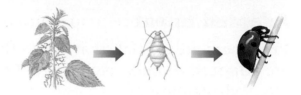

Food webs

Food chains can be joined together to make **food webs**. These show how some animals or plants are eaten by more than one type of animal.

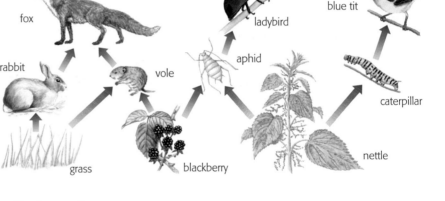

Plants are **producers** (they make the food that they need). Animals are **consumers** (they eat other organisms). Consumers can be split into three groups:

o **herbivores** only eat plants

o **carnivores** only eat meat

o **omnivores** can eat both meat and plants.

1 Look at the food web above. Which two animals eat blackberries?

2 These words have had their vowels removed. What should they say?
prdcr   prdtr   mnvr   crnvr

3 Think of something that you have eaten today. Draw a food chain that includes yourself and the food that you ate. Was the food that you ate a producer or a consumer?

Copy and complete using these words:

producers   food chains   omnivores
herbivores   carnivores

_____ _____ show who eats who.
Plants are _____ as they make the food that they need. Consumers eat other organisms. _____ eat only plants, _____ eat only meat, and _____ eat both plants and animals.

What have I learnt?

1 Copy these sentences into your book. You will need to unscramble the words in bold.

   a A **tibatah** is a place where plants and animals live.

   b The **nimrevonenlat steafreu** of a habitat determine which plants and animals will live there.

2 Decide whether each statement is true or false. Write them in your book, correcting the ones that are false.

   a A penguin has thick layers of fat to keep it warm. It is streamlined, which means that it can move through the water quickly.

   b A mole has good eyesight, and can sense smells and vibrations. It also has long front limbs for digging.

3 Name two animals that are adapted to cope with seasonal changes. Explain how they are adapted to cope with them.

4 Which of these animals are predators and which are prey?

   lion   snail   hare   spider
   heron   vulture

5 Use the following information to draw a food web:

   o Wood mice, aphids and caterpillars eat the leaves, acorns and buds from oak trees.

   o Great tits eat aphids, caterpillars and acorns.

   o Blue tits eat caterpillars and aphids.

   o Weasels and stoats both eat wood mice, great tits and blue tits.

   o Sparrowhawks eat small birds.

6 Imagine that you have been asked to write a short script for a nature programme called 'Down the garden path'. You are going to describe the animals that live in your back garden.

   o Describe the environmental conditions in the habitat.

   o Describe the animals and plants, and how they are adapted to live there.

   o Make your script interesting so that people will watch the programme.

Variation and classification

What I should already know

O There are differences between different living organisms.

O How to group organisms that are similar.

O How to recognise the features that help an organism survive in its habitat.

What I am going to meet in this unit

O How to identify similarities and differences between living organisms.

O Causes of inherited and environmental variation.

O How to sort organisms into two main groups, and then into smaller groups.

O The difference between vertebrates and invertebrates.

O Features of the five vertebrate groups.

O Features of the seven invertebrate groups.

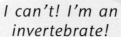

Come on, put your back into it!

I can't! I'm an invertebrate!

Species

Organisms of the same type are called a **species**. Members of the same species can **reproduce** with each other to produce **offspring** (babies) that are also able to reproduce.

The dogs in the picture have many features in common, but there are also some differences. These differences are called **variation**. Variation occurs between members of the same species.

These animals are both dogs. They are members of the same species.

Human beings also belong to the same species. Different people have different **hair** or **eye colours**, **heights** and **weights**. These are all examples of variation.

Look at the people in the picture. Can you find five examples of variation between them?

1 Unscramble these words:

  piesces   trivianoa   derporecu

2 Write a sentence for each of the words in question **1** to explain what they mean.

3 Look at the students in your class. What is the most common hair colour? What other hair colours do the students in your class have?

Copy and complete using these words:

**differences   reproduce   variation
species   offspring**

Organisms of the same type belong to the same _____. They can _____ with each other to produce _____ that can also reproduce. _____ between members of the same species are called _____.

Organisms of the same species can look different because of **variation**. There are two types of variation:

o Inherited variation
o Environmental variation

Inherited variation

A new baby is produced when a sperm fuses with an egg. Half of the **genetic information** for the baby comes from the father and half from the mother. This means that the child will have characteristics from both parents. Different children of the same parents usually inherit different characteristics. This is **inherited variation**.

Environmental variation

Environmental variation is caused by an organism's environment (where it lives). Factors like diet and illness can cause environmental variation in animals. Factors like how much water and light there is can cause environmental variation in plants.

Some characteristics like height are examples of both inherited and environmental variation.

Mum   Dad

Matthew   Rebecca

Rebecca looks similar to her brother Matthew, but there are also differences. They have inherited different characteristics from each parent.

1 What kind of information does a child inherit from its parents?

2 Look at the picture above. Write down two characteristics that Rebecca has inherited from her Mum and two that she has inherited from her Dad.

3 Which of the characteristics below are examples of environmental variation?

    eye colour   **a scar**   **permed hair**
    pierced ears   **large nose**

Copy and complete using these words:

inherited   environment   variation
both   characteristics

There are two different types of _____.
The _____ that we inherit from our parents are examples of _____ variation. Environmental variation is caused by an organism's _____.
Some characteristics are due to _____ types of variation.

Have you ever lost something because your bedroom was untidy? **Sorting** things into groups (tidying up) makes it easier to find them.

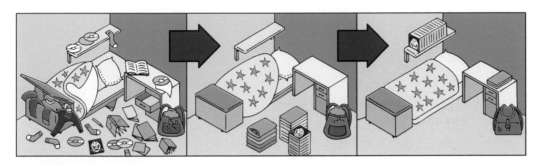

Classification

Scientists use a similar system for sorting living organisms. This system is called **classification**. Organisms are sorted into groups with the same characteristics.

An easy way to start classifying an organism is to decide whether it is a **plant** or an **animal**. Plants can then be sorted into two main groups depending on whether they are **flowering** or **non-flowering**. Animals can also be sorted into two main groups. Those that have a **backbone** and those that don't.

1 What does 'classification' mean?

2 These words have had their vowels removed. What should they say?

plnts   nmls   flwrng   bckbn

3 Sort these organisms into two groups, plants and animals. Which animals have a backbone?

**daisy   squirrel   salmon   grass
spider   jellyfish   dolphin**

Copy and complete using these words:

**plants   animals   classification
sorted   characteristics**

Organisms can be _____ into groups. They are grouped according to their _____. This system is called _____. The first step in classification is to sort organisms into two main groups, _____ and _____.

Animals can be sorted into two main groups. Animals that have a backbone are called **vertebrates**. Animals that don't have a backbone are called **invertebrates**.

Vertebrates

Vertebrates all have backbones but they can be very different. A bird is very different to a fish! Because there are so many different vertebrates, they are sorted into five smaller groups by their characteristics.

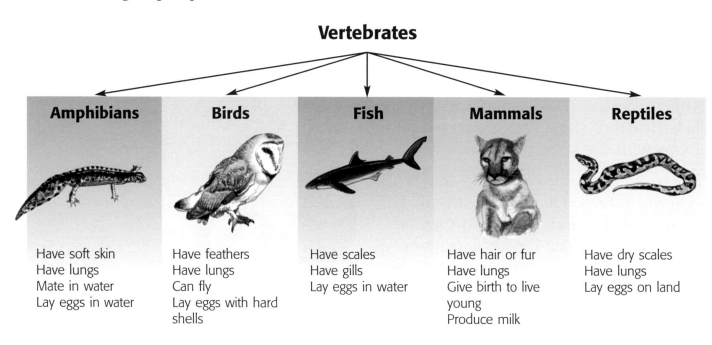

Vertebrates

Amphibians	Birds	Fish	Mammals	Reptiles
Have soft skin	Have feathers	Have scales	Have hair or fur	Have dry scales
Have lungs	Have lungs	Have gills	Have lungs	Have lungs
Mate in water	Can fly	Lay eggs in water	Give birth to live young	Lay eggs on land
Lay eggs in water	Lay eggs with hard shells		Produce milk	

1 What do we call an animal that does not have a backbone?

2 Name the five different vertebrate groups.

3 Which vertebrate group does this animal belong to?

'I have tough skin with hard, dry scales so that I can slither along the ground. I breathe using my lungs and I lay eggs on land.'

Copy and complete using these words:

mammals   vertebrates   fish amphibians   invertebrates

Animals with backbones are called _____. Animals without backbones are called _____.

Vertebrates can be sorted into five groups. These are _____, birds, _____, reptiles and _____.

Invertebrates

Animals that don't have a backbone are called **invertebrates**. They can be sorted into seven smaller groups.

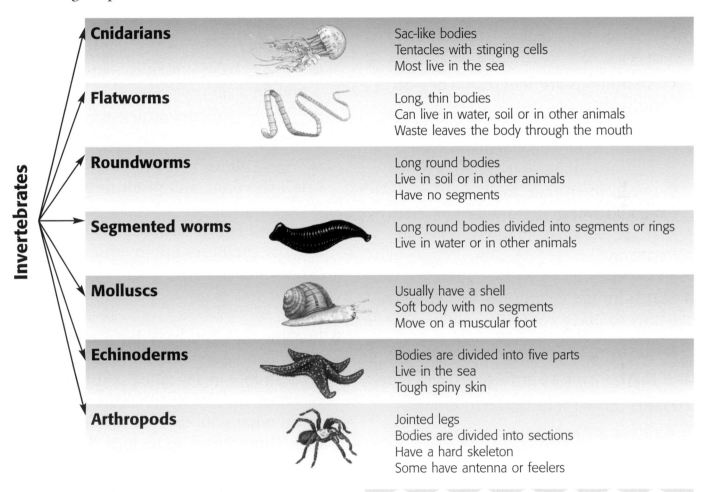

Invertebrates

Cnidarians		Sac-like bodies Tentacles with stinging cells Most live in the sea
Flatworms		Long, thin bodies Can live in water, soil or in other animals Waste leaves the body through the mouth
Roundworms		Long round bodies Live in soil or in other animals Have no segments
Segmented worms		Long round bodies divided into segments or rings Live in water or in other animals
Molluscs		Usually have a shell Soft body with no segments Move on a muscular foot
Echinoderms		Bodies are divided into five parts Live in the sea Tough spiny skin
Arthropods		Jointed legs Bodies are divided into sections Have a hard skeleton Some have antenna or feelers

1 How many invertebrate groups are there?

2 Which of the animals below is not an invertebrate?

    jellyfish   snail   horse   spider

3 Lobsters have hard skeletons and their bodies are divided into sections. Which invertebrate group do lobsters belong to? Explain your answer.

Copy and complete using these words:

invertebrates   molluscs
flatworms   segmented worms

Animals without backbones are called _____. Invertebrates can be sorted into seven groups. These are cnidarians, _____ _____, roundworms, _____, echinoderms and _____.

What have I learnt?

1 Decide whether each statement is true or false. Write them in your book, correcting the ones that are false.

  a Poodles and bulldogs are different species.

  b Giraffes and elephants are different species.

  c All humans are identical. They do not vary.

2 Look at the other students in your class. Write down five differences between them that are examples of inherited variation, and two differences that are examples of environmental variation.

3 Match up the beginnings and endings below to make complete sentences.

Beginnings

Sorting organisms into groups

Plants can be split into two groups,

Animals are split into two groups,

Endings

flowering and non-flowering.

is called classification.

those that have backbones and those that don't have backbones.

4 Copy and complete the sentences below. Remember to fill in the names of the vertebrate groups.

  a _____ have feathers and can fly.

  b _____ have scales, lay eggs and breathe through gills.

  c _____ have soft skin, mate in water and lay eggs.

  d _____ have dry scales, breathe through lungs and lay eggs on land.

  e _____ have hair or fur, give birth to live young and produce milk.

5 Draw and label a diagram to show a mollusc (molluscs are invertebrates). What characteristics does your mollusc have?

6 Imagine that you have been asked to write some questions for a new quiz programme on TV called 'Variation and classification'. Write five questions about what you have learnt in this unit. Don't forget to include the answers.

Acids and alkalis

What I should already know

O When a solid dissolves it is still there, but we can't see it.

O Different substances behave differently when they are added to water.

O Some changes are reversible (they can be turned back easily) and others are irreversible (they can't be turned back easily).

O Some hazards of burning.

What I am going to meet in this unit

O The names of some acids and alkalis.

O What acids and alkalis are.

O Some acids and alkalis are more dangerous than others.

O What indicators are.

O Some of the uses of acids and alkalis.

O Neutralisation.

O Different safety symbols.

Stop! Oranges have acid in them. Acids are really dangerous!

Its OK. Some acids are safe!

7E.1 Are all acids dangerous?

Acids in the home.

Examples of acids

Lots of substances around us contain acids. They have a sharp, tangy taste. We say that they taste sour. Acids in the home include things like **ethanoic acid** in vinegar and **citric acid** in lemons. Acids in the laboratory include **hydrochloric acid** (HCl), **sulphuric acid** (H_2SO_4) and **nitric acid** (HNO_3).

Being safe with acids

Not all acids are dangerous. Some acids are safe and do not harm us (weak acids). Others are very dangerous (strong acids). Dangerous chemicals have hazard symbols on them. They warn people about the risks of using them.

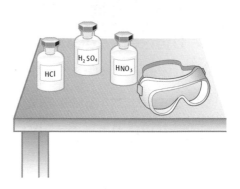
Acids in the laboratory.

 Corrosive: destroys living tissues like the eyes and skin.

 Irritant: not corrosive, but makes the skin go red or blister.

 Toxic: can cause death if swallowed, breathed in or absorbed through the skin.

 Harmful: can make you ill, but not as dangerous as toxic substances.

1 Name four different acids.

2 Unscramble these words and explain what they mean.

   nitrarti   rumlafh   sicoovrer

3 Lorries that are used to carry chemicals have hazard symbols on them. Why is this?

Copy and complete using these words:
chemicals hydrochloric hazard citric

_____ acid is found in fruit. _____ acid is used in the laboratory. Some _____ are dangerous. We use _____ symbols to warn people about the risks of using them.

Using indicators

Alkalis used in the home.

Alkalis

Alkalis are the chemical opposite of acids. Like weak acids, weak alkalis are not dangerous and are found in soaps and many other cleaning products. Like strong acids, strong alkalis can be corrosive. We say that corrosive alkalis are **caustic**. One of the most common alkalis is **sodium hydroxide**.

Telling the difference

Acids and alkalis are often colourless solutions. To tell them apart we use **indicators**. These are chemicals that **change colour** when they are mixed with an acid or an alkali.

Litmus is a natural indicator that is made from plants. It turns red in an acid and blue in an alkali. Litmus is very useful, but it only tells you if a substance is an acid or an alkali. It doesn't tell you how dangerous it is.

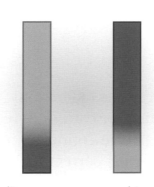

Blue litmus paper goes red in acid.
Red litmus paper goes blue in alkali.

1 What colour does litmus turn if it is mixed with an alkali?

2 These words have had their vowels removed. What should they say?

lkls   **cstc**   **ndctr**

3 Which of the substances below are alkalis?

toothpaste   **vinegar**   **tea**
lemon juice   **bleach**

Copy and complete using these words:

**indicators   caustic   blue
red   alkalis**

_____ are the chemical opposites of acids. They can be _____.

_____ turn one colour in an acid and another colour in an alkali. Litmus turns _____ in an acid and _____ in an alkali.

Universal indicator

Universal indicator is a special indicator that can tell you how **strong** an acid or an alkali is. It is a mixture of different indicators.

We measure the strength of acids and alkalis using the **pH scale**. Substances with a pH value of less than 7 are acids whilst those with a pH value of more than 7 are alkalis. A pH value of 7 is **neutral**.

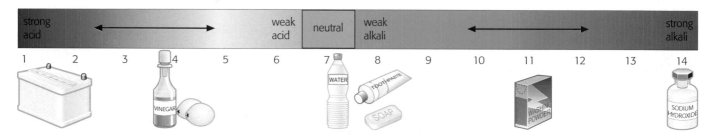

Measuring pH accurately

We can use information communications technology (ICT) to measure the pH of solutions more accurately than if we use an indicator. A **pH probe** connected to a data logger allows us to measure the pH correct to 1 decimal place.

1 What do we call the scale used to measure the strength of acids and alkalis?

2 These words have had their vowels removed. What should they say?

   dt lggr   nvrsl ndctr   strng

3 Which of the substances below is an alkali and which is neutral?

   washing powder (pH 11)
   water (pH 7)   vinegar (pH 5)

Copy and complete using these words:

strong   14   universal   pH   probe

_____ indicator tells us how _____ or weak an acid or an alkali is. The _____ scale runs from 0 to 14. 1 is a strong acid and _____ is a strong alkali. pH can be measured accurately using a data logger and a pH _____.

Neutralisation

Acids and alkalis are chemical **opposites**. When they are mixed together they cancel each other out. This is called a **neutralisation reaction**. They form a **neutral solution**. A neutral solution has a pH value of around 7.

An acid solution: there is more acid than alkali.

A neutral solution: there is the same amount of acid and alkali. They are balanced.

An alkaline solution: there is more alkali than acid.

Alkalis have a pH of more than 7. If acid is added to an alkali the pH gets lower. If enough acid is added the solution becomes neutral.

Acids have a pH of less than 7. If alkali is added to an acid the pH gets higher. If enough alkali is added the solution becomes neutral.

1 Unscramble these words:

rentula   **eislurantanoit**

2 What is the chemical opposite of an acid? What happens when the two are mixed together?

3 Bee stings are acidic. If you were stung by a bee would you treat it with an acid or an alkali? Explain your answer.

Copy and complete using these words:

mixing   **lower**   **alkali**   **neutral**

A neutral solution can be made by _____ an acid and an _____. A _____ solution has a pH of about 7.

When acid is added to an alkali, the pH gets _____.

Neutralisation reactions are important. We use them for many different things in everyday life.

Indigestion (stomach ache) can be caused by too much acid in the stomach. We can treat it by taking a medicine that is alkaline.

Bee stings hurt because they are acidic. We can treat them with an alkaline solution to neutralise the acid.

The pH of **soil** affects how well some crops will grow. If a soil is too acidic the farmer can add lime (finely ground limestone) to increase the soil's pH.

Bacteria in your mouth feed on left-over food and produce an acid that causes **tooth decay**. Brushing your teeth properly neutralises the acid because toothpast is alkaline.

1 Name two uses of neutralisation reactions in our everyday lives.

2 Name the substance that can be added to acidic soil to increase its pH.

3 Many lakes in countries like Sweden are affected by acid rain. They have become too acidic for organisms to live there. How would you treat them?

Copy and complete using these words:

neutralisation   **toothpaste**
tooth decay   **neutralises**

_____ reactions are useful in our everyday lives. For example, _____ is alkaline. Brushing our teeth properly _____ the acid produced by bacteria and helps to prevent _____ _____.

What have I learnt?

1 Match up the beginnings and endings below to make complete sentences.

Beginnings

Ethanoic acid is

Citric acid is

The formula for hydrochloric acid

A corrosive substance can

Endings

destroy living tissue.

found in oranges and lemons.

is HCl.

found in vinegar.

2 Decide whether each statement is true or false. Write them in your book, correcting the ones that are false.

a If red litmus paper is dipped into an acid it will stay red.

b If blue litmus paper is dipped into an acid it will stay blue.

c If red litmus paper is dipped into an alkali it will turn blue

d If blue litmus paper is dipped into alkali it will turn red.

3 Hydrochloric acid is added to a solution of sodium hydroxide. Just enough acid is added to neutralise the solution. What would you expect the pH of the solution to be? How could you prove this?

4 Copy the picture of the pH scale below. Decide what the pH of each substance is likely to be and write it under the pH scale in the correct place. The first one has been done for you.

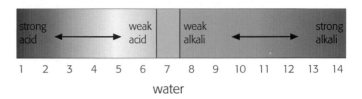

| strong acid | ← → | weak acid | weak alkali | ← → | strong alkali |

1 2 3 4 5 6 7 8 9 10 11 12 13 14
water

a Water
b Lemon juice
c Sodium hydroxide
d Toothpaste
e Hydrochloric acid

5 Explain why it wouldn't be a good idea to treat a bee sting on your arm with hydrochloric acid (pH 1). You should be able to think of at least two reasons.

6 Draw a line down the middle of your page. On the left of the line draw pictures of as many acids as you can think of. On the right of the line draw pictures of as many alkalis as you can think of.

Can you think of anything that is neutral that you could draw in the middle of your page?

Simple chemical reactions

What I should already know

- The names of some gases, like oxygen and carbon dioxide.

- Reversible changes can be turned back easily.

- Irreversible changes can't be turned back easily.

- Bubbles mean that a gas is being produced.

What I am going to meet in this unit

- A physical change does not make new substances.

- Air is needed for burning. The gas in air that is needed for burning is oxygen.

- New substances form during chemical reactions.

- What happens when metals react with carbonates, acids or oxygen.

- The fire triangle.

Ta daa!!

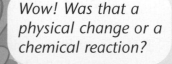

Wow! Was that a physical change or a chemical reaction?

And for my next trick ...

No new substances are made in a **physical change**. The change can be reversed easily. In a **chemical reaction** new substances are made. Chemical reactions can't be reversed easily.

More about chemical reactions

In a chemical reaction new substances are made. The substances that we start with are called the **reactants** (they react together). The new substances that are made are called the **products**. We can describe what happens in a chemical reaction by writing a **word equation**.

reactants ⟶ products

In this chemical reaction the product is a cake.

flour + baking powder + sugar + butter + eggs ⟶ cake

These signs can tell us that a chemical reaction is taking place:

o bubbles of gas are produced

o smoke is produced

o it feels warm

o the colour changes

o it glows (light is given out)

o a precipitate (solid) forms in a solution.

1 Name three signs that tell us that a chemical reaction has taken place.

2 Unscramble these words:

codstrup   **sactreata**

3 Write a word equation to describe how a cake is made.

Copy and complete using these words:

products   chemical   reactants

A _____ reaction can't be reversed easily. The substances that react are called the _____. The substances that are made are called the _____.

Acids are corrosive. They **corrode** some materials (eat away at them). **Corrosion** is an example of a chemical reaction. When some metals react with an acid, the pieces of metal get smaller or disappear, and bubbles of gas can be seen. A **salt** is produced.

Hydrochloric acid (HCl) reacting with a piece of magnesium ribbon.

Testing the gas

The gas that is produced when magnesium reacts with hydrochloric acid is **flammable** (it burns). If we hold a lit splint to the test tube during the reaction, we can hear a 'squeaky pop'. This shows that **hydrogen** is produced. We can write a word equation to describe the chemical reaction:

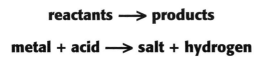

reactants ⟶ products

metal + acid ⟶ salt + hydrogen

Hydrogen is a colourless gas that doesn't smell. It is much lighter than oxygen.

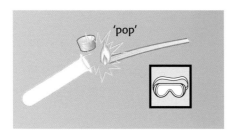

A lit splint makes a squeaky pop if hydrogen is present.

1 What is the name of the gas that is produced when magnesium reacts with hydrochloric acid?

2 What do we call the reaction when an acid eats away at something?

3 Write a short description of hydrogen. What test can we do to see if hydrogen is produced during a chemical reaction?

Copy and complete using these words:

corrosive   **chemical**
squeaky pop   **hydrogen**

Acids are corrosive. Corrosion is a type of _____ reaction. When some metals react with an acid, bubbles of gas are produced. A lit splint makes a _____ _____ if hydrogen is present.

Carbonates are substances that contain **carbon** and **oxygen**. They are found in baking powder and rocks like chalk. Carbonates fizz when they react with an acid because **carbon dioxide** gas is produced. We can write a word equation to describe the reaction:

acid + carbonate ⟶ salt + water + carbon dioxide

Testing for carbon dioxide

Like hydrogen, carbon dioxide is a colourless gas that doesn't smell. Carbon dioxide stops things burning. We can test for carbon dioxide by putting a lit splint into a test tube. The flame goes out if carbon dioxide is present.

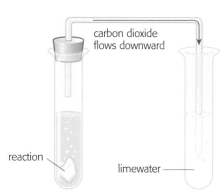

carbon dioxide flows downward

reaction

limewater

The limewater test. Carbon dioxide is heavier than oxygen and dissolves in water to produce an acid.

We can also use a liquid called **limewater** to test for carbon dioxide. When carbon dioxide is bubbled through it the liquid turns **cloudy** or milky white.

1 Write a word equation to describe the reaction between an acid and a carbonate.

2 These words have had their vowels removed. What should they say?

crbn   lmwtr   xygn   cldy

3 How is carbon dioxide similar to hydrogen? How is it different?

Copy and complete using these words:

**carbon dioxide   limewater
cloudy   water**

When carbonates react with an acid
_____ _____, a salt and _____ are produced. We can test for carbon dioxide using a lit splint or _____, which turns
_____.

Burning substances

Oxygen is one of the main gases in air. We need it to stay alive. It is also needed for things to burn.

Combustion (burning) is a chemical reaction that produces substances called **oxides**. When magnesium is burnt in air it reacts with oxygen to form magnesium oxide. We can write a word equation to describe the chemical reaction that happens when a metal like magnesium is burnt:

Oxygen is needed to burn things.

metal + oxygen —> metal oxide

The fire triangle

Three things are needed for something to burn: oxygen, heat and fuel. If one of these things is missing a fire will go out. To put out a fire you need to remove one of these things. If a burning chip pan is covered with a fire blanket, the fire can't get enough oxygen to keep burning. This puts the fire out.

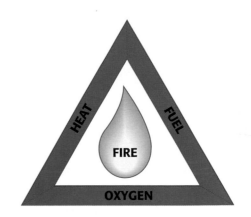

The fire triangle shows the three things needed for something to burn.

1 What is the name of the gas in air that is needed for things to burn?

2 Unscramble these words:

yengox   athe   lufe

3 When calcium is burnt in air a chemical reaction takes place. Write a word equation to describe this chemical reaction.

Copy and complete using these words:

oxide   combustion   oxygen   heat

Burning is also known as _____. When a metal is burnt in air, it reacts with _____ to produce a metal _____.

A fire will only burn if there is a fuel, _____ and oxygen.

Fuels are substances that store **chemical energy**. They can be burnt to produce **heat** and **light** energy.

- **Gas** can be used in ovens and camping stoves.
- **Crude oil** is used to produce petrol.
- **Coal** can be used in the home for fires.
- **Wax** is used to make candles to provide light.

Many fuels are **hydrocarbons** (substances made of **carbon** and **hydrogen**). When they burn they form two oxides. These are **carbon dioxide** and **water** (dihydrogen monoxide).

hydrocarbon + oxygen $\longrightarrow$ carbon dioxide + water + energy

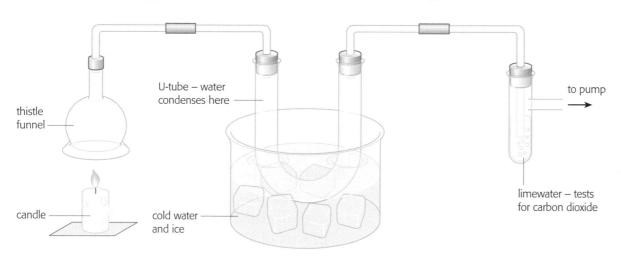

thistle funnel

U-tube – water condenses here

to pump

candle

cold water and ice

limewater – tests for carbon dioxide

You can use this apparatus to show that carbon dioxide and water are produced when hydrocarbons are burnt.

1 What do we call a substance that is burnt to produce energy?

2 These words have had their vowels removed. What should they say?

wx   cl   crd l   gs

3 A poisonous gas is formed if fuels don't have enough oxygen to burn properly. What is this gas called?

Copy and complete using these words:

carbon dioxide   chemical   water hydrocarbons

Fuels store _____ energy. Many fuels are _____. When hydrocarbons are burnt they produce _____ _____, _____ and energy.

1 Name six signs that can tell you that a chemical reaction has taken place. Describe what happens for each one.

2 Corrosion is the name for the chemical reaction that happens when a metal reacts with an acid. Draw the hazard symbol that is used to show that a substance is corrosive.

3 Match up the beginnings and endings below to make complete sentences.

Beginnings
Carbonates are substances that contain
When a carbonate reacts with an acid
We can use limewater to test for
Carbon dioxide is

Endings
a salt, water and carbon dioxide are produced.
carbon and oxygen.
colourless, has no smell and is heavier than oxygen.
carbon dioxide.

4 Copy and complete these word equations:

a _____ + _____ $\longrightarrow$ iron oxide

b magnesium + oxygen $\longrightarrow$ _____

5 Decide whether each statement is true or false. Write them in your book, correcting the ones that are false.

a A fuel stores elastic energy, and releases heat and light energy when it is burnt.

b Gas, oil and coal are all examples of fuels.

c A hydrocarbon contains carbon and oxygen.

d When a hydrocarbon is burnt, the reactants are carbon dioxide and water.

6 Design a leaflet about fire safety. Use the fire triangle in your leaflet to explain how a fire blanket can be used to put out a fire.

Particle model of solids, liquids and gases

What I should already know

○ How solids and liquids differ.

○ Water is ice that has melted. Ice is water that has frozen.

○ Some of the differences between solids, liquids and gases.

○ Changes of state can be changed back easily.

What I am going to meet in this unit

○ Some of the different properties of solids, liquids and gases.

○ All materials are made up of particles.

○ The particle theory of solids, liquids and gases.

○ How to explain the properties of solids, liquids and gases using the particle theory.

○ How theories are made.

Is it a solid, liquid or a gas?

Materials can be sorted into three groups: solids, liquids and gases. These groups are called the **three states of matter**. Solids, liquids and gases have different **properties**:

Solids

- are hard
- can't be poured
- are not easy to **compress** (squash)
- have a fixed shape and volume
- **conduct** (transfer) heat well

Liquids

- can be **poured**
- are not easy to compress (squash)
- take the shape of their container
- have a fixed volume
- conduct heat less well than solids

Gases

- are easy to compress (squash)
- take the shape of their container
- are less **dense** (heavy) than solids and liquids
- don't have a fixed shape or volume
- don't conduct heat well

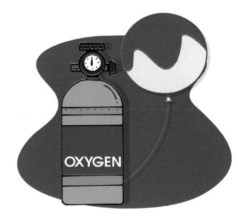

1 What are the three states of matter? How can we tell them apart?

2 Unscramble these words:

> dutconc   esdne   osprcmes

3 Desribe the properties of liquids using the words below:

> pour   container   squash
> volume   conduct

Copy and complete using these words:

**matter   compress   properties
poured   volume**

There are three states of _____. They have different _____. Solids have a fixed shape and _____. Liquids can be _____. Gases are easy to _____.

We can explain the different **properties** of solids, liquids and gases using the **particle theory**. Everything around us is made up of tiny bits called **particles**. Because the particles in solids, liquids and gases are arranged differently they have different properties:

The particles in **solids**…

o are **very close** together

o **vibrate** but can't move around

o are arranged in **rows**

o don't have much **energy**

The particles in **liquids**…

o are **close** together (most touch other particles)

o vibrate and can **move** past each other

o have more energy than the particles in a solid

The particles in gases…

o are quite **far apart** (you could fit several other particles between them)

o can move quickly in any **direction**

o have **lots of energy**

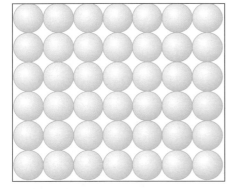
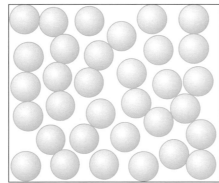
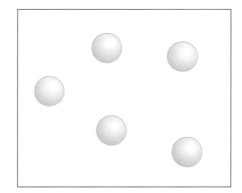

1 How can we explain the different properties of solids, liquids and gases?

2 Unscramble these words:

   yerv solec   bretavi   regney

3 Draw three pictures to show how the particles are arranged in solids, liquids and gases. How do the particles move in solids, liquids and gases?

Copy and complete using these words:

**past   particles   quickly
liquids   direction   vibrate**

The _____ in solids are very close together and _____. The particles in _____ are close together and can move _____ each other. The particles in gases move _____ in any _____.

Making theories

A **theory** (like the particle theory) is an explanation of how or why something happens. How are theories made?

The first step is to do **experiments**, or to **investigate** something. You may start with a **question** or a **prediction** (what you think will happen) to help you.

When you do your **experiments** you will make **observations** (what you see happening). You may also get some **results**.

You can then make a **conclusion** (what you think happened).

Your **theory** will be your explanation of how and why you got the results that you did. You can change your theory if you do some more experiments and make new conclusions.

The diamond was stolen last night...

We don't know how the thief got in. No windows were broken.

Hmm...

This is the security guard. He was here last night.

I didn't see anything strange.

Hmm... The security guard had a cut hand and there was blood on the broken glass. I'd like to see inside the guard's locker.

The diamond!

The guard used his key to get into the room. He cut his hand when he broke the glass case.

You're nicked!

1 What is a theory? Name an example of a theory.

2 Unscramble these words:

   ocnuiclsno **esvatorinbo**
   citridopen

3 In the cartoon above, what does the detective investigate? What are his observations? What does he conclude? What is his theory of what happened?

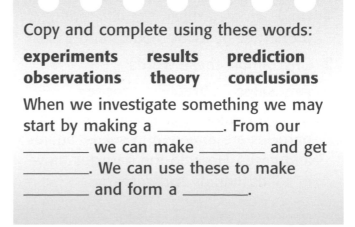

Copy and complete using these words:

**experiments   results   prediction
observations   theory   conclusions**

When we investigate something we may start by making a _____. From our _____ we can make _____ and get _____. We can use these to make _____ and form a _____.

Explaining properties

Changing state

In **changes of state** the particles don't change. The only things that change are the **energy** of the particles (how quickly they move) and the amount of **space** between them.

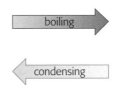

Compressibility

The particles in a solid are too close together to be **compressed** (squashed). Liquids can be compressed a little. Gases can be compressed a lot because there is lots of space between the particles.

solid   liquid   gas

Density

Density means how heavy something is for its size. **Dense** materials have lots of particles packed into a small space. Solids are denser than liquids and gases because they have lots of particles packed close together.

These blocks are the same size but one is heavier than the other. It is denser (has more particles packed into the same size space).

1 Why are gases more easily compressed than solids?

2 Copy the keywords below and explain what each one means:

compress   dense   melt evaporate

3 Describe what happens to the particles in an ice cube when it melts. What happens to the particles if the water is then boiled?

Copy and complete using these words:

particles   dense   space close   energy   compress

In changes of state the _____ of the particles and the _____ between them changes. Solids and liquids are hard to _____ because their particles are very _____ together. If a material is _____ it has a lot of _____ packed into a small space.

Diffusion

Diffusion happens when particles spread out and mix without being stirred. Solids can't **diffuse** because their particles are too close together for other particles to move between them. Liquids and gases are able to diffuse. Gases diffuse more quickly than liquids because their particles are able to move quickly in different directions.

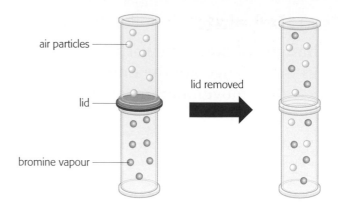

When the lid separating the two gas jars is removed, the particles spread out and mix.

Pressure

Gas particles move very quickly in all directions. **Gas pressure** is caused when gas particles **collide** with (bump into) the sides of a container.

The pressure on the outside of an open can is the same as the pressure inside it, so it keeps its shape. If the air is sucked out of the can it collapses. The pressure on the outside of the can is now larger than on the inside.

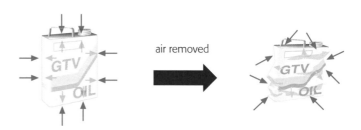

The can collapses when the air is sucked out of it.

1 Draw a diagram like the one above to show what happens when particles diffuse.

2 These words have had their vowels removed. What should they say?

prssr   dffsn   clld

3 Why does a can collapse if the air is sucked out of it?

Copy and complete using these words:

**gases   pressure   container
diffusion   collide   liquids**

_____ happens when particles mix and spread out. _____ and _____ are able to diffuse. Gas _____ is caused when gas particles _____ with the walls of a _____.

What have I learnt?

1 Decide whether each of the products below is a solid, a liquid or a gas.

a lemonade

b butter

c tomato sauce

d jelly

2 Write down the properties of each of the three states of matter (solid, liquid and gas).

3 Match up the beginnings and endings below to make complete sentences.

Beginnings

In a solid
In a liquid
In a gas

Endings

the particles are far apart and have lots of energy.
the particles are very close together and vibrate.
the particles are close together and can move past each other.

4 Decide whether each statement is true or false. Write them in your book, correcting the ones that are false.

a Solids can be compressed because their particles are very close together.

b Gases are very dense because lots of particles are packed into a small space.

c The particles in a liquid have less energy than the particles in a gas.

5 Draw particle diagrams to explain why solids can't diffuse but gases can.

6 Write a detective story like the one on page 52. Make it clear what the detective investigates, observes and concludes. Use the questions below to help you.

o What is the detective investigating?

o What does the detective observe?

o What is the detective's conclusion?

o What theory does the detective make to explain what happened?

Solutions

What I should already know

- Sieves separate solids from liquids.
- Some solids are soluble, others are insoluble.
- When a solid dissolves we can't see it, but it is still there.
- When distilled water evaporates nothing is left because it is pure.
- Steam from salty water is not salty. The water evaporates, but the salt is left behind.

What I am going to meet in this unit

- How to get a solid out of a solution.
- Factors that affect how quickly a solid dissolves.
- Soluble solids and insoluble solids.
- How to test liquids to see if they are pure.
- How to get pure water from sea water.
- How to separate the colours in an ink.

I'm so thirsty! If only I could get pure water from sea water.

Shame he doesn't know about distillation!

Chemistry Set

Some substances **dissolve** in liquids like water. We say that they are **soluble**. When a substance dissolves it seems to disappear, but it is still there. Its particles mix evenly with the particles of the liquid to form a **clear solution**. Substances that do not dissolve in a liquid are **insoluble**.

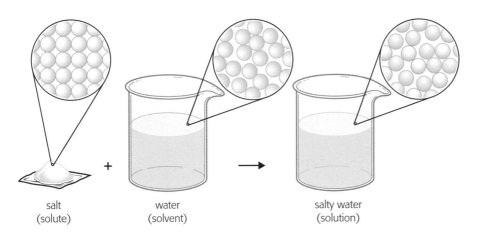

salt
(solute)

water
(solvent)

salty water
(solution)

Water is not the only solvent. Nail varnish is insoluble in water but is soluble in propanone (found in nail varnish remover).

A substance that dissolves in a liquid is called a **solute**. Salt dissolves in water so salt is a solute. The liquid that a solute dissolves in is called a **solvent**. Water is a solvent. The mixture that forms (salty water) is called a **solution**. A solution is a mixture of a solute and a solvent.

1 What word describes a substance that dissolves in a liquid?

2 Unscramble these words:

oletnsv   usilotno   esltno

3 Make a list of substances that are soluble in water. When you have done that, make a list of substances that are insoluble in water.

Copy and complete using these words:

solvent   solution   soluble   solute

A _____ substance dissolves in a liquid. The substance is called a _____. The liquid that it dissolves in is called a _____. The mixture that forms is called a _____.

Separating mixtures

We can use **separation techniques** to separate mixtures. The method used depends on the mixture.

Filtration

This method separates an **insoluble solid** from a liquid. The **filtrate** (liquid) goes through the paper but the **residue** (insoluble solid) does not.

Sand and water can be separated by filtration.

Evaporation

This method separates a **soluble solid** (solute) from a solvent. The solvent **evaporates** when the solution is boiled.

When seawater is boiled the water evaporates, leaving solid salt behind.

Chromatography

This method separates more than one soluble substance from a solvent. The substances travel across a piece of paper with the solvent. Some substances are more soluble than others. The substances travel different distances across the paper, depending on how soluble they are.

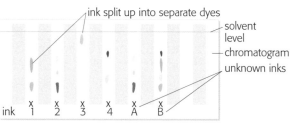

Inks are made up of different dyes. These can be separated by chromatography. Unknown inks can be identified.

1 What do we call the solid that is left after a mixture has been filtered?

2 How could you separate the different dyes in an ink?

3 Imagine that you have cooked some rice. How could you separate the rice from the water? Explain how this method will work.

Copy and complete using these words:

**evaporation   soluble   filtration
chromatography   solvent   liquid**

_____ separates an insoluble solid from a _____. _____ separates a _____ solid from a liquid. _____ separates more than one soluble substance from a _____.

When we use evaporation to separate a solute from a solvent, the solute is left behind but the solvent escapes. If we want to collect the solvent we need to use a technique called **distillation**. We can use distillation to collect more than one solvent from a mixture.

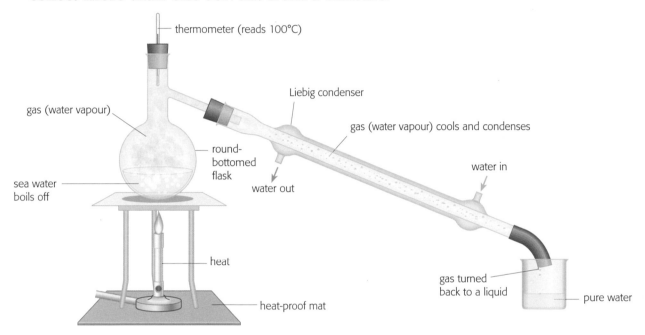

thermometer (reads 100°C)

gas (water vapour)

Liebig condenser

gas (water vapour) cools and condenses

round-bottomed flask

water in

sea water boils off

water out

water in

heat

gas turned back to a liquid

pure water

heat-proof mat

We can get pure water from sea water by distillation. The sea water is boiled so that the water evaporates. The water vapour cools in the **condenser** and **condenses** back to liquid water. The pure water can then be collected.

1 Unscramble these words:

nesdcneos   sevraoepat   ilatilidsnot

2 What technique would you use to get pure water from sea water?

3 Imagine that you are on a desert island with only some matches, a funnel, two empty tins and a rubber tube. How could you extract pure water from sea water?

Copy and complete using these words:

**evaporates   sea   collected
condenser   pure**

We can use distillation to get _____ water from _____ water. The sea water is heated until the water _____. The water vapour condenses back into a liquid in the _____ and is _____.

The **solubility** of a substance tells us how much of it will dissolve in a solvent. Solubility is affected by the factors below:

o how quickly the solution is **stirred**

o the **size** of the bits of solute

o the **solvent** that the solute is dissolved in

o the temperature of the solvent.

The solubility of a substance increases as the temperature increases. More of the substance will dissolve if the solvent is hot than if it is cold. We have to say what the temperature is when we talk about the solubility of a substance. For example: 'Potassium chloride has a solubility of 37.6 g/ 100 g of water at 30 °C.'

Sugar dissolves faster in hot tea than in cold tea.

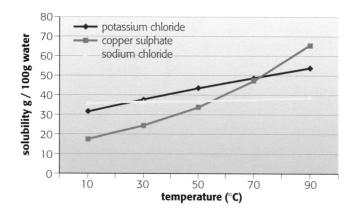

Solubility curves for three different substances. At certain temperatures some substances are more soluble than others.

1 What four factors can affect the solubility of a solute?

2 These words have had their vowels removed. What should they say?

tmprtr   **slblt**   **strrd**

3 Use the solubility curves above to find the solubility of copper sulphate at 50 °C. What is the solubility of copper sulphate at 90 °C?

Copy and complete using these words:

**solvent   temperature   substance
solubility   increases**

The _____ of a substance tells us how much of it will dissolve in a _____.
The solubility of a substance _____ as the _____ increases. We have to say what the temperature is when we talk about the solubility of a _____.

Saturated solutions

When a solute dissolves, the solvent particles mix with the solute particles. The solute particles move in between the solvent particles. If too much solute is added, there are too many solute particles to fit in between the solvent particles. Not all of the solute can dissolve. We say that the solution is **saturated**. The solute **sinks** to the bottom of the beaker.

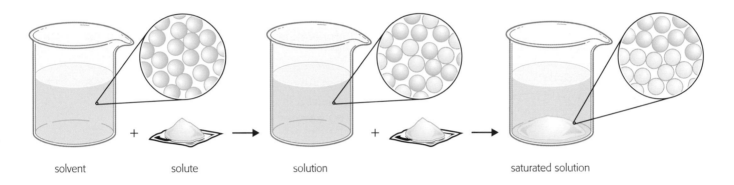

solvent   solute   solution   saturated solution

A **saturated solution** is a solution in which no more solute can dissolve at that temperature. The solubility of potassium chloride is 37.6 g/ 100 g of water at 30 °C. This tells us that we will make a saturated solution if we add more than 37.6 g to 100 g of water at 30 °C.

1 Describe what happens to the solute particles when a solution becomes saturated.

2 Unscramble these words:

seisldov   ardtasute

3 Copper sulphate has a solubility of 24.2 g/ 100 g of water at 30 °C. How could you make a saturated solution of copper sulphate at 30 °C?

Copy and complete using these words:

**solute   solvent   temperature
solubility   saturated   sinks**

A _____ solution is made when no more solute can dissolve in the _____ at that _____. The solute _____ to the bottom of the beaker. We can make a saturated solution if we know the _____ of the _____.

What have I learnt?

1 Which of the solids below is the odd one out? Explain your answer.

chalk   sand   sugar   sawdust

2 Copy the diagram below and add these labels to it:

filtrate   **conical flask**

beaker   **residue**

mixture of sand and water

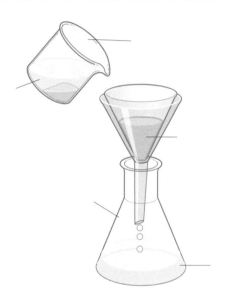

3 Sort these sentences into the correct order to describe distillation.

o The gas condenses into a liquid and can be collected.

o The solution is heated until the solvent evaporates.

o We can separate and collect a solvent from a solute by distillation.

o The gas is cooled in the condenser.

4 Which separation technique would you use if you wanted to separate the different dyes in an ink?

5 Use the data in the table below to plot a solubility curve for a substance called supa-solid (remember to label the axes). How does the solubility of supa-solid change as the temperature of the water increases?

Temperature (°C)	0	20	40	60	80	100
Mass (g) dissolved in 100 g water	29	37	46	55	66	77

6 Imagine that you are a forensic scientist. A strange liquid was found at a crime scene and the police want you to investigate it.

Write a plan of the experiments that you could do to identify what is in the liquid. Remember to write down what you would expect to find out from each experiment.

Energy resources

What I should already know

- Plants need sunlight to make their food.
- Burning is a chemical reaction. New substances are made.
- Insulation helps to keep something warm or cold.
- Animals need food so that they can grow.

What I am going to meet in this unit

- Energy is released when fuels are burnt.
- Fossil fuels are non-renewable energy resources.
- Some energy resources are renewable.
- Ways that we can save energy.
- Energy can be transferred.
- Living things transfer energy.

Typical fossil fuel. They never last.

Where did Gas go?

He ran out!

Fuels are substances that release (give out) **energy** when they burn. Energy is needed to make things happen. Fuels release **light** and **heat** energy when they burn.

We use **Bunsen burners** in science lessons to heat things. The fuel that is burnt in a Bunsen burner is called **natural gas**. When we light a Bunsen burner, we must wear safety goggles and tie back long hair. We should then follow these safety rules:

1 Put the Bunsen on a heat-proof mat.

2 Close the Bunsen's valve (air hole).

3 Turn the gas tap to half open.

4 Light the Bunsen with a splint. You can then open or close the valve to change the flame.

5 Keep the valve closed (gives a yellow flame) when you are not heating something.

cool yellow flame

hot blue flame

valve closed

valve open

Use for safety   Use for heating

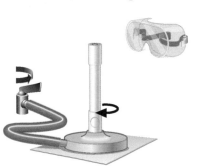  

1 Unscramble these words:
   thigl   athe   ngyeer

2 What colour Bunsen burner flame is used for safety, and what colour flame is used to heat things?

3 Draw and label a diagram of a Bunsen burner. What are the safety rules that you must follow when using a Bunsen burner?

Copy and complete using these words:

**natural gas   blue   energy
yellow   fuels**

_____ release _____ when they are burnt. Bunsen burners burn a fuel called _____ _____. When the valve is closed the flame is _____. This is used for safety. We heat things with a _____ flame.

Coal, oil and natural gas are energy resources that are known as **fossil fuels**. They formed from the remains of plants and animals that lived on Earth over 300 million years ago. They store **chemical energy** and release heat and light energy when they burn. We can use them to make **electricity**.

How were fossil fuels formed?

When tiny sea creatures died they were buried under layers of rock. Heat and pressure eventually changed them into oil and gas. Coal was formed in a similar way when trees died and were buried under layers of rock.

1 Tiny sea creatures died and sank to the sea floor.

2 They were covered by layers of sand and mud which became layers of rock. Heat and pressure turned their remains into oil and gas.

3 The gas and oil were squeezed through tiny holes in the rocks until they reached a layer that they couldn't pass through. We can now extract them and use them as fuels.

Fossil fuels take millions of years to form. We can't easily make some more. We say that they are **non-renewable**. They will eventually run out.

1 Name three fossil fuels. What two types of energy do these fossil fuels release when they burn?

2 What do we mean when we say that fossil fuels are non-renewable?

3 Think of two things that oil is used for as a fuel. Try to do the same for coal and then for gas.

Copy and complete using these words:

oil   plants   fossil   non-renewable

Coal, _____ and natural gas are _____ fuels. They formed from the remains of _____ and animals over 300 million years ago. Fossil fuels are _____. They will eventually run out.

Saving energy

Why save energy?

Fossil fuels are **non-renewable**. This means that they will eventually run out. The more we use fossil fuels for energy, the faster they will run out. If we use less energy, the fossil fuels will not run out as quickly. **Saving** energy (using less energy) is important because it means that there will be fossil fuels for us to use for longer.

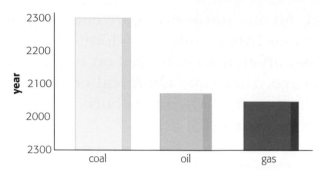

This bar chart shows when we think the three main fossil fuels will run out.

These are examples of ways in which we can save energy:

o Turn lights off when you leave a room

o Cycle or walk to school

o Turn the heating down at home

o Use low-energy light bulbs in your home

o Stop heat from escaping from draughty doors and windows

o **Insulate** your loft (this stops heat escaping)

o Have double glazed windows fitted

A low-energy light bulb.

1 Look at the bar chart at the top of the page. Which of the three fossil fuels are we likely to run out of first? When do we think this will happen?

2 Explain why we should all try to save energy.

3 Write a list of things that you could do to help save energy. Try to think of some examples that aren't on the list above.

Copy and complete using these words:

**cycling   save   lights
non-renewable**

Fossil fuels are _____ and will eventually run out. We need to _____ energy to make fossil fuels last longer. We can all save energy by walking or _____ to school, and switching off _____ when we are not using them.

71.4 Renewable energy

Not all of the energy we use is from non-renewable fossil fuels. There are now other ways to **generate** (make) **electricity** that can be replaced and will not run out. These are called **renewable** energy resources.

Types of renewable energy

Energy resource	How it works
Wind power	Wind turbines are linked to a generator (makes electricity).
Wave power	Wave movement drives a turbine linked to a generator.
Hydroelectric power	Water runs downhill and drives a turbine linked to a generator.
Solar panels	Energy from the Sun heats water that can be used in the house.
Biofuel (biomass)	Plant and animal waste is used to make methane gas which is burnt, or wood is burnt.
Geothermal power	Hot rocks under the Earth's surface are used to make steam. This is used to drive turbines and generators.

Solar panels use energy from the Sun to heat water.

1 These words have had their vowels removed. What should they say?

   bmss   wnd   gthrml

2 Which renewable energy resources use moving water to generate electricity?

3 How are renewable energy resources different to non-renewable ones? Is it better to use renewable resources or non-renewable resources? Explain your answer.

Copy and complete using these words:

solar   generate   renewable
geothermal   wave   resources

_____ energy resources can be replaced and will not run out. They can be used to _____ electricity. Wind power, _____ power, hydroelectric power, _____ panels, biofuel and _____ power are renewable energy _____.

Using energy

How do we get our energy?

The Sun is the source of almost all the energy resources on Earth. Energy from the Sun can be **transferred** into other types of energy.

Plants use **light energy** from the Sun to make their food. This involves a chemical reaction called **photosynthesis**. Animals can't make their own food. Food is their **fuel**. Food contains **chemical energy**. Animals eat food to get the energy that they need to stay alive. This involves a chemical reaction called **respiration**.

Nutrition Information

	○ Typical value per 100 g		● 30 g serving with 125 ml of semi-skimmed milk	
ENERGY	1650 kJ	390 kcal	750 kJ*	180 kcal
PROTEIN	6 g		6 g	
CARBOHYDRATES	83 g		31 g	
of which sugars	35 g		17 g	
starch	48 g		14 g	
FAT	3.5 g		3 g*	
of which saturates	0.7 g		1.5 g	
FIBRE	2.5 g		0.8 g	
SODIUM	0.65 g		0.25 g	
VITAMINS:		(% RDA)		(%RDA)
THIAMIN (B₁)	1.2 mg	(85)	0.4 mg	(30)
RIBOFLAVIN (B₂)	1.3 mg	(85)	0.6 mg	(40)
NIACIN	15 mg	(85)	4.6 mg	(25)
VITAMIN B₆	1.7 mg	(85)	0.6 mg	(30)
FOLIC ACID	333 μg	(165)	110 μg	(55)
VITAMIN B₁₂	0.85 μg	(85)	0.75 μg	(75)
IRON	7.9 mg	(55)	2.4 mg	(17)

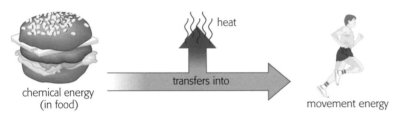

chemical energy (in food)   heat   transfers into   movement energy

We can find out how much energy there is in a food by reading the label on the package. Energy is measured in **joules** (J). 1 **kilojoule** is 1000 joules.

The label on this breakfast cereal box tells you how much energy is in the food.

1 What is the source of almost all the energy resources on Earth?

2 Unscramble these words:

lejosu   jokesilluo

3 Imagine that you have just eaten a bowl of cornflakes. Explain how the energy that you have got from eating the bowl of cornflakes originally came from the Sun.

Copy and complete using these words:

energy   chemical   light   Sun

The _____ is the source of almost all the energy resources on Earth. Plants need _____ energy from the Sun to make their food. Food contains _____ energy. Animals eat food to get the _____ that they need.

What have I learnt?

1 Put the sentences below into the right order to give safety instructions for lighting a Bunsen burner.

- o Turn the gas tap to half open.
- o Put your safety goggles on.
- o Close the valve (air hole).
- o Light the Bunsen with a splint.
- o Put the Bunsen on a heat-proof mat.

2 Decide whether each statement is true or false. Write them in your book, correcting the ones that are false.

a Non-renewable energy resources will not run out. They can be replaced.

b Coal, oil and wind power are examples of non-renewable energy resources.

c Oil and gas were formed from the remains of tiny sea creatures that were buried under layers of rock.

3 Explain how walking or cycling to school is a way of saving energy.

4 Which of the energy resources below are renewable resources? Which are non-renewable resources?

solar panels   **coal**

wind power   **hydroelectric power**

natural gas   **geothermal power**

wave power   **biofuel**

5 Why is eating food like putting petrol into a car?

6 Write an advert for a new car that uses water as a fuel. You should explain why this new car is better than a car that uses petrol as a fuel.

Use the words and phrases below in your advert:

- o fuel
- o energy resources
- o renewable
- o non-renewable
- o save energy

Electrical circuits

What I should already know

- The dangers of electricity.
- How to identify materials that conduct electricity.
- Cables and wires are good conductors of electricity.
- How to build a circuit.
- How to include a switch in a circuit.

What I am going to meet in this unit

- How to build simple circuits, and draw circuit diagrams.
- What current is, and how it is measured.
- How switches in a circuit affect current.
- Series and parallel circuits.
- Adding bulbs to a circuit affects the current.
- How to be safe when using electricity.

I thought you said that we were going to do circuit training.

We are!

What is a circuit?

Many of the things we use every day, like televisions and kettles, use **electrical circuits**. To work, **electric charge** (electricity) must be able to flow around these circuits. The flow of electric charge is called the **current**.

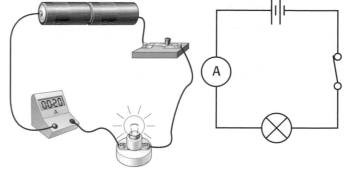

We can draw **circuit diagrams** to show how a circuit works. Circuit diagrams use special symbols to show what the **components** (parts) are. Some of these symbols are shown below:

Component	Circuit symbol	What it does
cell		Provides the current that flows around the circuit.
bulb (filament lamp)		**Transforms** (changes) electrical energy into heat and light energy.
switch	or	Breaks the circuit and stops the current flowing when open. Closes the circuit so that the current can flow when closed.
ammeter		Measures the current in **amps** (A).

1 Which component measures the current in a circuit?

2 These words have had their vowels removed. What should they say?
bttry   swtch   crct

3 Draw a circuit diagram that shows a circuit made up of a cell, a bulb and a switch. The switch should be open.

Copy and complete using these words:

**current   circuit   diagram
electric charge**

An electrical circuit works when an _____ _____ flows around it. The flow of electric charge is called the _____. We can draw a circuit _____ to show how a _____ works.

An **electrical device** like a kettle will only work if the circuit is **complete**. The current must be able to flow around all of the circuit. If there are any breaks in the circuit the current can't flow and the device won't work.

In this circuit one of the cells is the wrong way around.

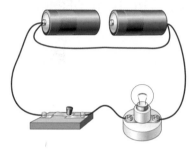

Wires have been **connected** to one side of both cells. For current to flow, different wires should be connected to both ends of the cell.

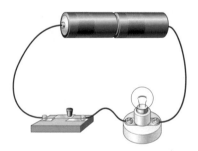

In this circuit the switch is open. This breaks the circuit and the current can't flow.

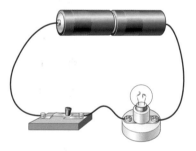

Here, the **filament** inside the bulb (a thin piece of wire) has snapped. The current can't flow. The bulb must be replaced.

1 What happens if there is a break in a circuit?

2 Why won't a bulb light if the filament inside it has snapped?

3 Draw a circuit diagram to show a circuit with two cells, two bulbs and a switch. Remember to draw the switch so that the bulbs will light.

Copy and complete using these words:

current   flow   device   circuit

An electrical _____ will only work if the _____ can flow all around the circuit. If there is a break in the _____ the current can't _____ and the device won't work.

7J.3 Adding more bulbs

The flow of electric charge around a circuit is known as **current**. We can measure the current in a circuit with an **ammeter**. Ammeters measure current in **amps** (A).

The circuit below is a **series** circuit (a simple loop). The current does not change as it flows around this type of circuit. The current is not used up by the bulb.

It does not matter where an ammeter is placed in a series circuit, the reading will be the same. The current does not change as it flows around the circuit.

Adding more bulbs

It is hard for current to flow through a bulb. Bulbs **resist** the flow of current. The more bulbs there are in a circuit, the harder it is for the current to flow.

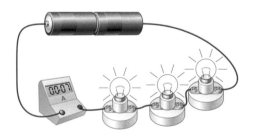

There are more bulbs in the circuit on the right so there is more resistance. The ammeter reading is lower and the bulbs are not as bright.

1 Which component is used to measure current in a circuit?

2 Unscramble these words:

   spam   setsri   temaerm

3 Does an ammeter give a higher or lower reading if you add more bulbs to a circuit? Use the words below to explain why:

   current   resistance

Copy and complete using these words:

**bulb   amps   resists
ammeter   series**

Current is measured in _____ (A) using an ammeter. The current is the same all around a _____ circuit. It doesn't matter were the _____ is placed. The current is not used up by a _____ but it _____ the flow of the current.

Series and parallel circuits

There are two types of electrical circuit. They are known as **series** and **parallel** circuits.

Series circuits

In a series circuit the components are in a single **loop**. You can trace your finger all the way around the circuit without taking your finger off. If there is a break or a gap in a series circuit it will not work because the current can't flow.

If there is a break in this series circuit all the bulbs will go out because the current can't flow.

Parallel circuits

In a parallel circuit the components are on different **branches**. The current flows around each branch. If there is a break in the circuit on one branch, the current can still flow around the circuit along the other branches. The components in the other branches will still work.

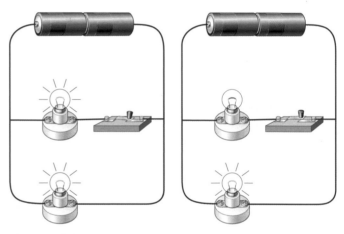

This parallel circuit has two branches. If there is a break in one of the branches the bulb on that branch goes out. The current can still flow around the circuit so the bulb on the second branch stays lit

1 These words have had their vowels removed. What should they say?

srs   prlll   brnchs

2 What happens if there is a break or a gap in one branch of a parallel circuit?

3 Draw circuit diagrams of the series and parallel circuits on this page. Remember to draw the switch as closed in the parallel circuit, so that the current will flow.

Copy and complete using these words:

**will   will not   branches
parallel   loop   series**

In a _____ circuit all of the components are in one _____. If there is a break in the circuit it _____ _____ work. In a _____ circuit the components are on different _____. If there is a break in one branch the components in the other branches _____ work.

The electricity we use at home carries far more energy than cells or batteries. We have to be very careful when we use **electrical devices**.

To protect us from harm, plugs on electrical devices are fitted with **fuses**. A fuse is a thin piece of wire. If too much electricity goes into a device the fuse in the **plug** melts. This breaks the circuit and the current can't flow. The device stops working.

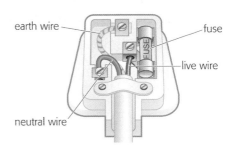

earth wire   fuse   live wire   neutral wire

A three-pin plug.

Safety rules

Here are some simple rules for using electrical devices safely:

Don't have too many plugs in a socket.

Don't use electrical devices in the bathroom. Water conducts electricity.

Don't use a device with frayed wires. Replace them.

Don't push things into sockets.

1 These words have had their vowels removed. What should they say?

fss   **sckt**   **dvc**   **plg**

2 What are fuses and how do they work? Why are plugs on electrical devices fitted with fuses?

3 Try to think of another rule for using electrical devices safely. Draw a safety picture like the ones above to go with your rule.

Copy and complete using these words:

breaks   fuses   electricity
devices   wire

Plugs on electrical devices have _____ to keep us safe. A fuse is a thin piece of _____ that melts and _____ the circuit if too much _____ goes into the device. We must be careful when using electrical _____.

What have I learnt?

1 Draw the circuit symbol for each of the components below:

   a An open switch.

   b An ammeter.

   c A bulb.

   d A cell.

2 What happens to the current if the filament inside a bulb in a series circuit breaks?

3 Match up the beginnings and endings to make complete sentences:

   Beginnings

   Current is measured in

   A bulb resists the flow of current but

   The greater the number of bulbs in a circuit

   Endings

   does not use it up.

   the harder it is for current to flow.

   amps (A) using an ammeter.

4 Draw a circuit diagram of a parallel circuit with two branches. There should be two bulbs on one branch, and a switch and bulb on the other branch.

5 Look at the picture below. Make a list of the electrical hazards that you can see. What could you do to make each hazard safe?

6 Imagine that you work on a TV quiz programme. Write ten questions for the programme about the things that you have met in this unit. Write them in the same style as the example below and remember to include the answers!

   Q: What A is the name of the piece of equipment used to measure the amount of current in a circuit?

   A: Ammeter.

Forces and their effects

What I should already know

- Forcemeters measure forces in Newtons.
- Different surfaces affect the speed of objects because of friction.
- Gravity, weight, air resistance and water resistance are types of force.
- Upthrust acts on objects in water.
- Elastic bands stretch when forces are added.

What I am going to meet in this unit

- Different types of force.
- Forces affect the speed and direction of objects.
- Friction tries to stop two surfaces from moving over each other.
- Upthrust pushes up on objects in liquids.
- Weight pulls down on objects.

I'm not sure that's what he meant when he said he wanted to 'feel the force'...

Forces are all around us. Although we can't see them we can see what they do. Forces can have these effects on an object:

- change its **shape**
- change its **speed**
- change its **direction**.

Different types of force

A force is a push or a pull. **Contact** forces act when things touch. Examples of contact forces are:

- upthrust
- water resistance
- friction
- air resistance (drag)

The forces acting on this parachute are slowing the person's fall to Earth.

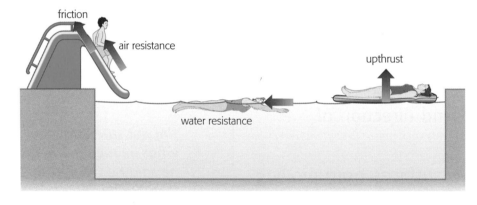

All of the forces shown in this picture are examples of contact forces.

Other forces act from a distance. These are called **non-contact** forces. Magnetism, static and gravity are examples of non-contact forces.

1 What are the three effects that forces can have on an object?

2 Unscramble these words:

   trintofic   shurtput   traygiv   catsit

3 Make a list of contact and non-contact forces. How are contact forces different to non-contact forces?

Copy and complete using these words:

**contact   direction   non-contact
shape**

Forces can change the speed, _____ or _____ of an object. _____ forces act when things touch. _____ forces act at a distance.

When an object floats it is because of two forces called **weight** and **upthrust**. Weight is a force that pulls the object down. Upthrust is a force that pushes the object up. When the two forces are the same we say that they are **balanced**. An object floats when weight and upthrust are balanced.

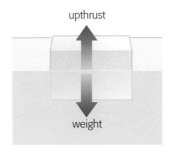

Upthrust keeps things afloat.

When you float in water, you feel as though you weigh less than you do out of the water. This is because the upthrust of the water cancels out some of the downward force of the weight.

An object will only float if its **density** is less than or equal to the density of water. Density means how heavy something is for its size.

the 20N object only weighs 12N

8N of upthrust

This object weighs 20N in air, but only 12N in water. There is an upthrust of 8N.

1 What are the names of the two forces that act on a floating object?

2 These words have had their vowels removed. What should they say?

wght   pthrst   dnsty   blncd

3 An object weighs 15N in air but only 9N in water. What is the force of upthrust that is acting on the object?

Copy and complete using these words:

**upthrust   balanced   two
weight**

When an object floats _____ forces act on it. _____ is the force that pulls the object down. _____ is the force that pushes the object up. The two forces must be _____ for the object to float.

We can use a **forcemeter** to measure the size of a force. A forcemeter contains a spring. Springs are **elastic**. They stretch (get longer) when you add a force to them, but go back to their original shape if you take the force away.

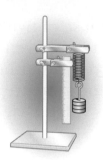

We can see how a force affects the length of a spring by adding weights to it and measuring the length of the spring.

Results from a stretching experiment.

Force (N)	Stretch (mm)
0	0
2	16
4	30
6	62
8	64
10	82
12	96
14	112

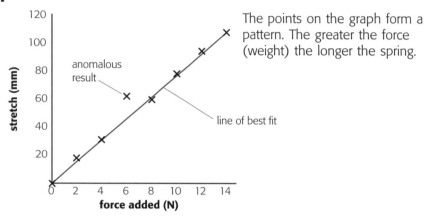

The points on the graph form a pattern. The greater the force (weight) the longer the spring.

A **line of best fit** is drawn through as many points as possible to help us see a **pattern**. One of the results on the graph doesn't sit on the line of best fit. It is an **anomalous** (unusual) result. To make the results more **reliable** we should repeat the experiment and work out the **average** value of each reading.

1 Unscramble these words:

**catlies   trapnet   labeleri
treemorecf**

2 How do we draw a line of best fit? What does it help us see?

3 Look at the results from the stretching experiment. How could you make the results from this experiment more reliable?

Copy and complete using these words:

**pattern   reliable   elastic
forcemeters   anomalous**

_____ measure the size of forces. They contain springs, which are _____. A line of best fit helps us see a _____. If we have an _____ result we should repeat the experiment to make it more _____.

Without friction we would slip over.

Friction is a force that acts when two surfaces move over each other. It tries to stop the movement of the two surfaces. Friction is useful because it stops things slipping. It makes things grip.

Friction isn't always useful though. It slows down the movement of things and can make them get hot. Tyres and parts of machines wear out because of friction. We can reduce the friction between two surfaces by using a **lubricant** to stop the two surfaces rubbing together.

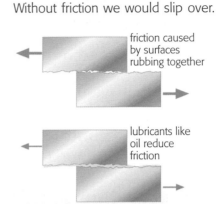

friction caused by surfaces rubbing together

lubricants like oil reduce friction

Air and water resistance

Air resistance is a kind of friction that acts when an object moves through air. **Water resistance** acts when an object moves through water. Smooth or dart-shaped objects move through air and water with less resistance than block-shaped objects. We say that they are **streamlined**.

This plane is streamlined to reduce the effect of air resistance.

1 Give one reason why friction can be useful, and one reason why it isn't always useful.

2 These words have had their vowels removed. What should they say?

frctn   **r rsstnc**   **strmlnng**

3 Draw a picture of a car that has been streamlined to reduce the effect of air resistance. What are the features of the car that make it streamlined?

Copy and complete using these words:

**air   surfaces   hot   streamlined
water   friction**

_____ is a force that acts when two _____ move over each other. Friction can be useful but it can also slow things down and make them get _____. We can reduce _____ resistance and _____ resistance by making objects _____.

81

Speed tells us how far an object has travelled and how long it took. We work it out by dividing the **distance** travelled by the **time** it took to travel that distance.

$$\textbf{speed} \; = \; \textbf{distance} \; \div \; \textbf{time}$$

The unit that we use for speed depends upon the units used for distance and time. We usually measure speed in **metres per second** (m/s). The distance is measured in metres (m). The time taken is measured in seconds (s).

Distance/time graphs

Distance/time graphs can tell us how fast something is moving. The slope of the graph shows the speed. The **steeper** the slope the **faster** the object is moving.

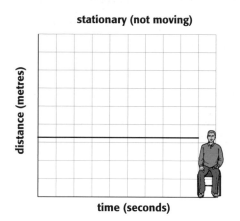

stationary (not moving)

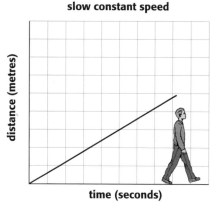

slow constant speed

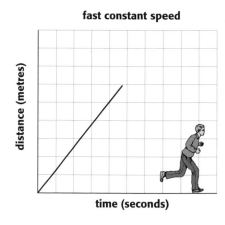

fast constant speed

1 How do you work out the speed of an object?

2 Which unit do we usually use when we measure speed?

3 If an object travels 10 metres in 4 seconds what is its speed? The object is travelling at a constant speed. Draw a distance/time graph to show this.

Copy and complete using these words:

dividing far distance long metres

Speed tells us how _____ an object has travelled and how _____ it took. We work it out by _____ the _____ travelled by the time taken. We usually measure speed in _____ per second.

What have I learnt?

1 Match up the beginnings and endings below to make complete sentences:

Beginnings

Forces can

Forces that act when things touch are called

Forces that act at a distance are called

Static and magnetism are

Endings

contact forces.

examples of non-contact forces.

change the shape, speed and direction of an object.

non-contact forces.

2 Draw a picture of a rubber duck floating in water. Add force arrows to show the forces acting on the rubber duck to keep it afloat (remember to label them).

3 The table below shows some results from a stretching experiment:

Weight added (N)	Stretch (mm)
1	50
2	58
3	70
4	74
5	82
6	90
7	102
8	125

Plot a line graph of these results (remember to label the axes). Draw a line of best fit (this should go through as many points as possible). Circle any points on the graph that you think are anomalous results.

4 Look at the list of examples below. Friction acts in each one. For each example decide whether the effect of friction is useful or a nuisance.

a Brakes stopping a car from moving.

b Part of a machine getting hot and wearing away.

c A parachutist falling to earth (remember that air-resistance is a type of friction).

d Tying a tie.

e An Olympic swimmer trying to break the world record (remember that water-resistance is a type of friction).

5 Draw two distance/time graphs. One should show an object that is stationary. The other should show an object that is moving at a fast constant speed

6 Choose ten important words from this topic and write them in a list. Write a sentence for each word to explain what it means.

The Solar System and beyond

What I should already know

- The Sun, Moon and Earth are almost spherical.
- The Earth spins on its axis every 24 hours.
- It is daytime when our part of the Earth faces the Sun.
- The Sun rises in the East and sets in the West.
- It takes a year for the Earth to orbit the Sun, and 28 days for the Moon to orbit the Earth.

What I am going to meet in this unit

- The Moon orbits the Earth.
- The Earth spins as it orbits the Sun.
- Why the Sun seems to move across the sky during the day.
- How we see the Sun and the Moon.
- How the Earth is different to the other planets in our Solar System.

*When I said that we were dressing up as stars, I meant **luminous** stars!*

Day and night

The Earth is a sphere that **orbits** (moves round) the Sun. It takes a **year** (about 365 days) to complete a full orbit. During the day the Sun seems to move across the sky in a curve. This is because the Earth is **rotating** (spinning) on its axis. It also makes the stars look as if they are moving across the sky at night. It takes a **day** (24 hours) to make a complete turn.

The Sun rises in the East and sets in the West.

When the part of the Earth that you are on faces the Sun it is day-time. When the part of the Earth that you are on no longer faces the Sun it is night-time.

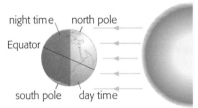

The Moon

The Moon is a **natural satellite** that orbits the Earth. It takes about 28 days to complete a full orbit. This is known as a **lunar** (Moon) month.

1 How long does it take the Earth to make a complete turn on its axis? How long does it take the Earth to orbit the Sun?

2 These words have had their vowels removed. What should they say?

rth   xs   dy   stllt

3 If it is day-time in Britain, is it day or night on the other side of the Earth? Explain your answer.

Copy and complete using these words:

rotates	**hours**	**orbits**
lunar	**day-time**	**days**

The Earth _____ the Sun and _____ on its axis. It takes 24 _____ to make a complete turn. It is _____ when our part of the Earth faces the Sun. The Moon takes 28 _____ to orbit the Earth. This is a _____ month.

The seasons

We have **seasons** because the Earth is tilted on its axis. The Earth is divided into two **hemispheres** (halves). The **northern hemisphere** is the top half of the Earth. The **southern hemisphere** is the bottom half of the Earth. We live in the northern hemisphere.

When the northern hemisphere is tilted towards the Sun, more of it is in sunlight than in darkness. This season is called **summer**. Days are longer and nights are shorter. When the northern hemisphere is tilted away from the Sun, more of it is in darkness than in sunlight. This season is called **winter**. Days are shorter and nights are longer.

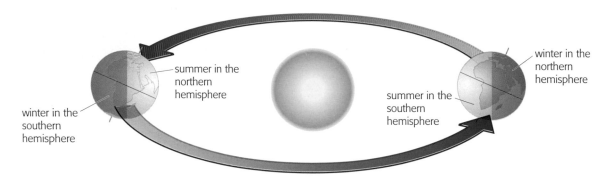

When the northern hemisphere is tilted towards the Sun it is summer in Britain. It is winter in the southern hemisphere.

1 What is the name given to the two halves of the Earth? Which half of the Earth do we live in?

2 Unscramble these words:

raye   **mumsre**   **tenriw**

3 Days are longer and nights are shorter in the summer. How does this change in the winter? Try to explain why this is.

Copy and complete using these words:

away   **hemisphere**   **seasons**
summer   **winter**   **axis**   **towards**

The Earth is tilted on its _____. This is why we have _____. We live in the northern _____. When the northern hemisphere is tilted _____ the Sun it is _____. When it is tilted _____ from the Sun it is _____.

The Sun gives out its own light. It is a **light source**. We say that it is **luminous**. The Earth and Moon do not give out light. They are **non-luminous**. We can see the Moon because it **reflects** light from the Sun.

Luminous objects.

Eclipses

When the Moon passes between the Earth and the Sun, it casts a shadow on the Earth. The part of the Earth where the shadow falls gets dark and the Sun seems to be blocked out. This is called a **solar eclipse**. When the Earth passes between the Sun and the Moon, the Earth casts a shadow on the Moon. This is called a **lunar eclipse**.

Non-luminous objects.

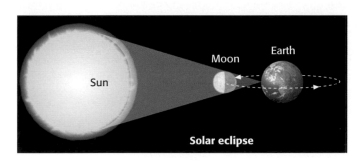

Solar eclipse

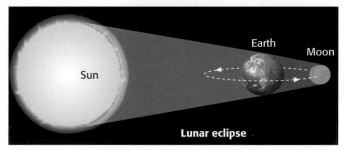

Lunar eclipse

1 Give two examples of luminous objects and two examples of non-luminous objects.

2 What do we mean when we say that the Moon is non-luminous? How do we see it?

3 Describe what happens when there is a solar eclipse. How is this different to what happens when there is a lunar eclipse?

Copy and complete using these words:

Moon   luminous   Earth   reflects

The Sun is _____. The Moon is non-luminous. We can see the Moon because it _____ light from the Sun. A solar eclipse happens when the _____ passes between the Earth and the Sun. A lunar eclipse happens when the _____ passes between the Moon and the Sun.

The Earth is one of nine **planets** in our **Solar System**. The Solar System also contains **asteroids, natural satellites** (moons) and the Sun (a **star**). The planets in our Solar System orbit the Sun.

Distances are not to scale

Planet	Distance from the Sun (million km)	Diameter of planet (km)	Time to spin on its axis (hours)	Time to orbit the Sun (Earth years)	Gravity (compared to Earth)	Average temperature (°C)
Mercury	58	4900	1416	0.25	0.4	350
Venus	108	12100	5832	0.65	0.1	480
Earth	150	12700	23.9	1.0	1	20
Mars	228	6800	24.6	1.9	0.4	-23
Jupiter	780	143000	9.8	12	2.6	-150
Saturn	1430	120000	10.2	29	1.2	-190
Uranus	2800	50000	10.8	84	1.1	-210
Neptune	4500	49000	15.8	165	1.4	-220
Pluto	5900	2400	153.6	248	0.08	-240

The conditions on the other planets are different to those on Earth.

1 Unscramble these words:

recrumy   tripuje   teennup   topul

2 On which planet is gravity the strongest?

3 How does the distance of a planet from the Sun affect the time that it takes to complete a full orbit?

Copy and complete using these words:

Earth   star   Sun   planets

There are nine _____ in our Solar System. They all orbit a _____ called the _____ The conditions on each planet are different to those on _____.

Our **Solar System** is made up of **planets**, **moons** (natural satellites), **asteroids** and a **star** called the Sun. Asteroids are big pieces of rock that come in different shapes and sizes. Most of the asteroids in our Solar system orbit the Sun in the asteroid belt between Mars and Jupiter. Moons are bigger than asteroids and orbit the planets. Planets are bigger than moons. The planets orbit the Sun.

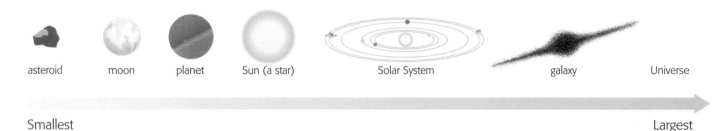

asteroid   moon   planet   Sun (a star)   Solar System   galaxy   Universe

Smallest   Largest

Although the Solar System is very big it is not as big as a **galaxy**. Galaxies are made up of millions of stars like the Sun. The Sun is part of a galaxy called the **Milky Way**. The **Universe** is made up of millions of galaxies like the Milky Way and is bigger than we can imagine. There is even evidence that the Universe is getting bigger!

1 What is the name of our galaxy? Is it bigger or smaller than our Solar System

2 What is an asteroid? Is an asteroid bigger or smaller than a moon?

3 Write a list of things that are bigger than the Earth (remember that some of the planets in our Solar System are bigger than the Earth).

Copy and complete using these words:

**Universe   asteroids   stars
planets   galaxies**

The _____ is made up of millions of _____, which are made up of millions of _____. Our star is called the Sun. The Solar System is made up of the Sun, _____, moons and _____.

What have I learnt?

1 Match up the beginnings and endings below to make complete sentences:

Beginnings
It takes 24 hours
The Earth is tilted
The Sun rises in
The Moon is

Endings
the East.
on its axis.
a natural satellite.
for the Earth to rotate on its axis.

2 Draw and label a diagram to show how the Earth is tilted towards the Sun when it is summer in the southern hemisphere.

3 Explain what happens during a lunar eclipse.

4 The mnemonic below helps us to remember the order of the planets from the Sun:

My	**M**ercury
Very	**V**enus
Easy	**E**arth
Method	**M**ars
Just	**J**upiter
Speeds	**S**aturn
Up	**U**ranus
Naming	**N**eptune
Planets	**P**luto

Make up your own mnemonic to help you remember the order of the planets from the Sun.

5 There are no photographs of the Universe. Why do you think this is?

6 Imagine that you have gone to another planet in our Solar System for a holiday. Write a postcard to a friend telling them what the planet is like. Use the questions below to help you write your postcard.

o Is the planet smaller or bigger than Earth?

o How long does a day last?

o How long does a year last?

o Is it hotter or colder on the planet than on the Earth?

o What are the neighbouring planets?

o Have you seen any little green men?

Glossary

Words in italic have their own glossary entry.

A

acid A substance with a *pH* lower than 7.

adapted Special features that help an organism survive in its *habitat*.

adolescence The period of change when a child develops into an adult.

air resistance A *force* that acts against an object moving through air.

alkali A substance with a *pH* higher than 7.

ammeter Measures electric *current* in amps (A) or milliamps (mA).

anomalous result A result that doesn't sit on the line of best fit. It is unusual.

B

boiling Changing a liquid into a gas by heating it.

C

carbon dioxide (CO₂) A colourless gas that stops things burning. Turns *limewater* cloudy.

carnivore An animal that only eats meat.

caustic substance A substance that is *corrosive* (usually an *alkali*).

cell The smallest unit that living things are made up of.

cell division The process by which our bodies grow and replace damaged *cells*.

cell membrane A thin skin that controls what goes into and out of a *cell*.

cell wall The outer layer of a plant *cell* that gives it its shape.

cervix A ring of muscle at the bottom of the *uterus*.

chloroplast The part of a plant *cell* that traps light energy so that the plant can make its own food. It contains chlorophyll.

chromatography A way of separating more than one soluble substance from a *solvent*.

classification The system used to sort living organisms into groups.

combustion Where a substance like a *fuel* reacts with *oxygen* to give out heat and light energy. Also called burning.

condensation The process by which a gas turns into a liquid.

consumer A living organism that eats other plants and animals.

corrosive substance A substance like an *acid* that reacts with or eats away at other materials. They can destroy living *tissues*.

current The flow of electric charge in a circuit, measured in amps (A).

cytoplasm The area inside a *cell* that is not the *nucleus* or a *vacuole*.

D

density How heavy something is for its size.

diffusion When *particles* spread out and mix without being stirred.

dissolving When a *solute* mixes with a *solvent* to make a *solution*.

distillation A separation technique that involves *evaporation* and *condensation*. The *solvent* can be collected.

E

eclipse When the Sun's light on the Moon or the Earth is blocked.

egg The female sex cell in animals.

embryo A young organism that is developing in an egg, or inside its mother.

evaporation The process by which a liquid changes into a gas. A way of separating a soluble *solute* from a *solvent*.

F

fertilisation When a male and a female sex cell join to make a new living organism.

fetus A developing human baby, nine weeks after *fertilisation*.

filtration A way of separating an *insoluble* solid from a liquid.

food chain A diagram that shows us who eats who.

food web A diagram that shows us how all the *food chains* in a community join together.

force A push or a pull, measured in newtons (N).

freezing The process by which a liquid changes into a solid.

friction A *force* that acts when two surfaces move over each other.

fuels Substances that store chemical energy. They produce heat and light energy when they are burnt.

fuse A safety device fitted into plugs on electrical devices.

G

galaxy A group of stars and their *solar systems*.

gas pressure This is caused when gas *particles* collide with the sides of a container.

H

habitat A place where plants and animals live.

herbivore An animal that only eats plants.

hibernation The way in which some animals survive the winter by sleeping through it.

hydrogen A light, colourless gas produced when some metals react with *acids*. Makes a squeaky pop when a lit splint is held to a test tube of it.

I

indicator A substance that changes colour when an *acid* or an *alkali* is present.

insoluble A substance that does not dissolve in a *solvent*.

invertebrate An animal that doesn't have a backbone.

J

joule (J) The unit for measuring energy (1 kJ = 1000 J).

L

limewater A liquid that turns cloudy or milky white when *carbon dioxide* is present.

litmus A natural indicator that turns red when an *acid* is present and blue when an *alkali* is present.

luminous object An object that is a light source.

M

melting The process by which a solid changes into a liquid.

menstrual cycle The monthly cycle that starts with the release of a ripe *egg cell* from an *ovary*, and ends with a period.

migration A regular journey made by some animals to breed or find food.

Milky Way The *galaxy* that our *Solar System* belongs to.

N

neutral solution A *solution* with a *pH* of 7. It is not acidic or alkaline.

nocturnal An animal that is active at night.

non-renewable resource An energy resource that will eventually 'run out'.

nucleus The part of a *cell* that controls what the cell does.

O

omnivore An animal that eats both plants and meat.

organ Part of a living organism with a special job, made up of different *tissues* working together.

organ system Many *organs* working together to do a special job.

ovary The *organ* that makes the female sex cells (*egg cells*).

oviduct The tube that the *egg cell* travels through to get to the *uterus*.

oxygen A colourless gas found in air that is needed for respiration and *combustion*.

P

parallel circuit An electric circuit with branches.

particles Tiny bits that make up everything around us.

penis Where the *sperm cells* leave the man's body.

pH scale A scale that is used to measure acidity and alkalinity. *Acids* have a *pH* of less than 7, *alkalis* have a *pH* of more than 7.

placenta The *organ* in female mammals that provides a *fetus* with food and *oxygen* and gets rid of waste.

predator On organism that hunts *prey* for food.

prey An organism that is hunted by a *predator*.

producer A green plant that makes its own food.

puberty The part of *adolescence* when we become sexually mature.

R

renewable resource An energy resource that can be replaced and will not 'run out'.

reproduction The process by which new living organisms are made.

S

saturated solution A *solution* in which no more *solute* can dissolve at that temperature.

series circuit An electric circuit with branches.

Solar System The Sun (a star) and the planets, moons and asteroids that orbit it.

solubility The amount of *solute* that will dissolve in a known volume of *solvent*.

solute The substance that dissolves in a *solvent*.

solution The mixture that forms when a *solute* dissolves in a *solvent*.

solvent The liquid that a solute dissolves in to form a *solution*.

specialised cells *Cells* that are *adapted* to do a certain job.

species Organisms of the same type that can *reproduce* with each other to produce fertile offspring.

speed This tells us how far an object has travelled and how long it took.

sperm cell The male sex cell in animals.

states of matter The three possible forms of materials – solids, liquids or gases.

T

theory An explanation of how or why something happens.

tissue A group of *cells* of the same type working together to do a job.

U

umbilical cord The tube that joins a *fetus* to the placenta.

universal indicator A mixture of different *indicators* that gives a certain colour at a certain *pH*.

Universe Everything that exists, including all of the *galaxies*.

upthrust A *force* that pushes an object upwards.

uterus An *organ* in the female body where a *fetus* develops.

V

vacuole A liquid-filled bag in plant *cells* that stores important chemicals.

vagina The part of the female reproductive system where *sperm* enter the female's body.

variation Differences between members of the same *species*.

vertebrate An animal that has a backbone.

W

water resistance A *force* that acts against an object moving through water.

weight A *force* that pulls an object downwards.

Index